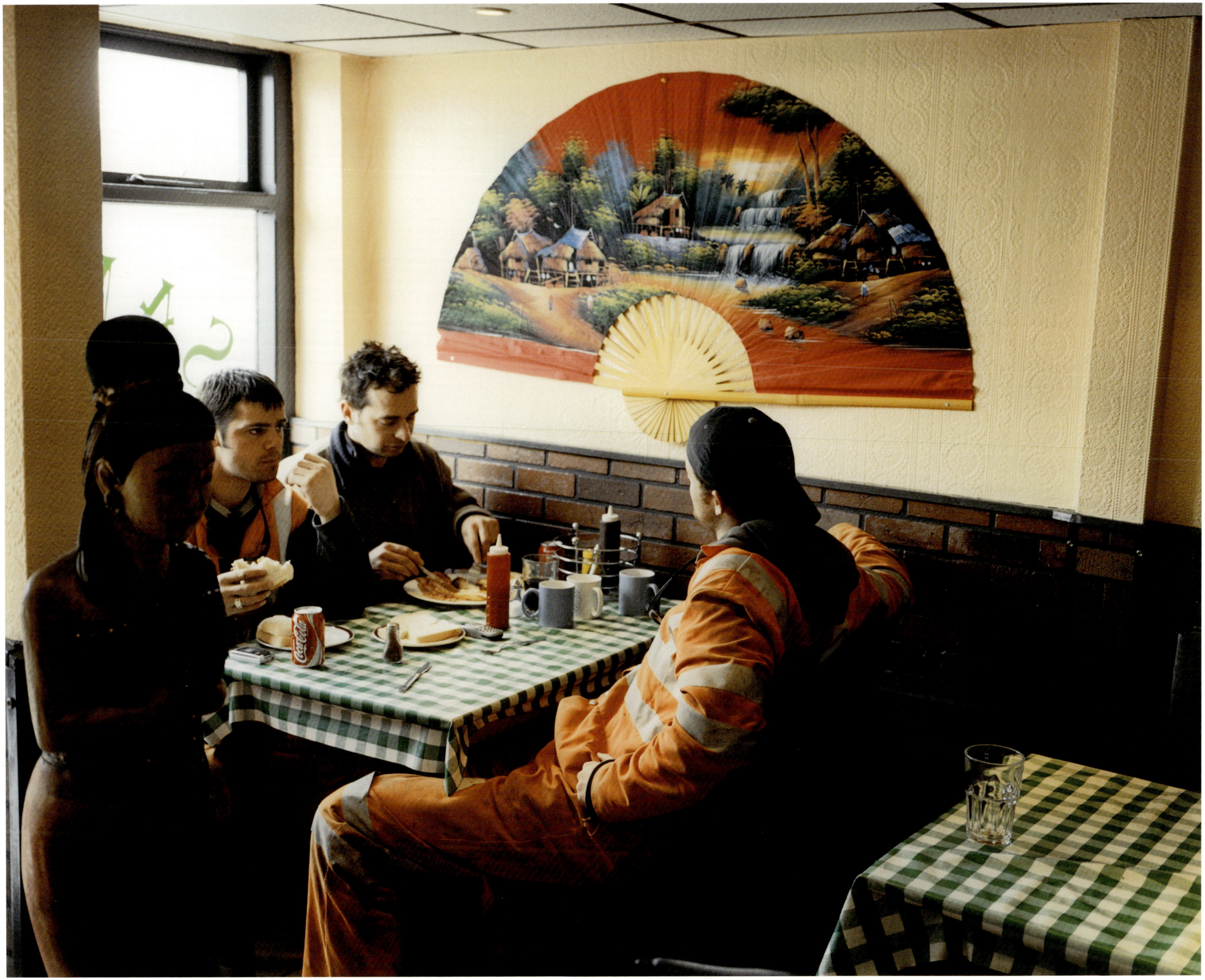
Coca-Cola

To Paul

All the best, from

Stephen

London, Jan 2005

# Invisible

**Stephen Gill**

**Published by Nobody**

These photographs show us a paradox: how fluorescent safety-clothing, which is intended to make workers in potentially dangerous occupations conspicuous, can make its wearers invisible. They are the latest series by a photographer with a sustained interest in the way cities work. Here he shows us the invisible people who ensure that London and its suburbs function. It is as if the metropolis is a complicated clock which no one ever sees – or rather notices – being wound up.

Photography has been well described by John Szarkowski as, at a basic level, equivalent to the act of pointing. Stephen Gill points and then points again and continues pointing until he has a collection of photographs of whatever phenomenon initially caught his attention. One of his first series addressed the urban motif of the shopping trolley. I must have seen those trolleys but I didn't really. Since Stephen showed me his portraits of (mostly) women pushing their shopping trolleys through London's streets, I have noticed them everywhere and in all their variety. His award-winning book *Field Studies* (2004) shows us other series which make various urban sights, like advertising hoardings seen from the back, jump into visibility. His photographs turn inside out (sometimes literally) or re-edit our habitual take on city life, introducing us – as if for the first time – to what our eyes have seen but apparently decided not to pass on to our minds.

Because Stephen works in series based on an idea, Martin Parr – one of his champions – has called him a conceptual photographer. This is so, in a sense, but not precisely in the sense of the Conceptual Art of around 1970. No doubt Invisible has affinities with Ed Ruscha's fabulous little books of photographs of unacknowledged Americana of the 1960s, such as gas stations, swimming pools and parking lots – but there is a difference. Stephen's subjects are pretexts for picture-making. Each of his subjects – whether safety-clothing, shopping trolleys, lost tourists reading maps or safety-barriers around road-works – gives him a pretext for picture-making. Each of his pictures here has the primary subject of safety-clothing – but then includes much more. His task is to make pictures and he does this constantly, methodically and yet inventively rather in the manner of Lee Friedlander, whose stylistic traces can be seen in some of these photographs, and (again, like Friedlander) with something of the absorbed craftsmanship of a shoe-maker. Part of the craft is to fill the photograph with small signs of the present, as if the photographer is an astute archaeologist of everyday life from the future. Another part is to make the photographer disappear. Actually, by sometimes wearing a fluorescent jacket and mounting his 6x7cm camera on a tripod, Stephen made himself invisible to most of his subjects. However, there is another kind of invisibility involved. I think Stephen likes the idea of 'ego-less' or vernacular photographs. In this little collection he manages to point us towards something that we had seen but perhaps never really noticed – while making himself almost invisible.

**Mark Haworth-Booth**

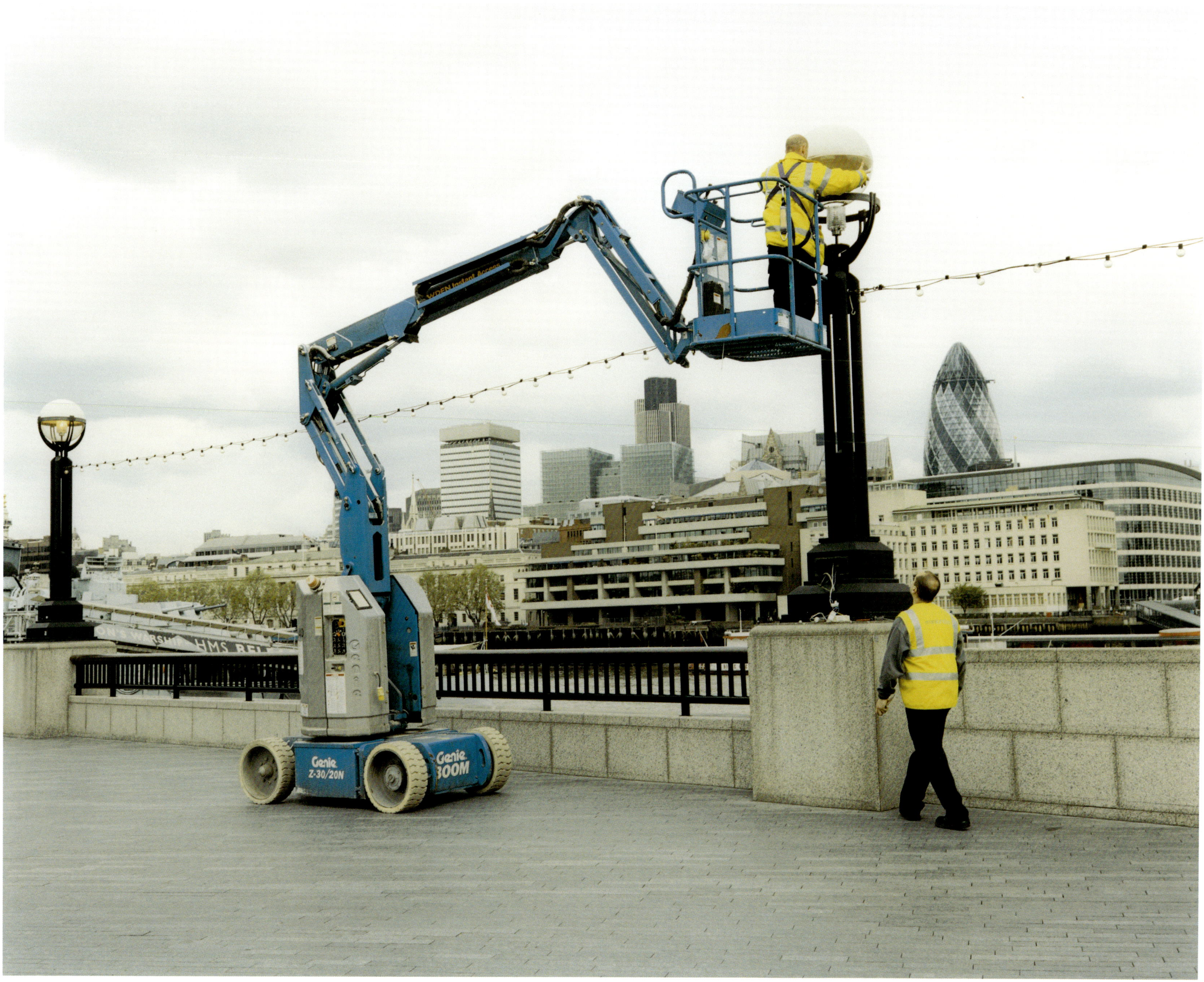
Genie
Z-30/20N
Genie
BOOM

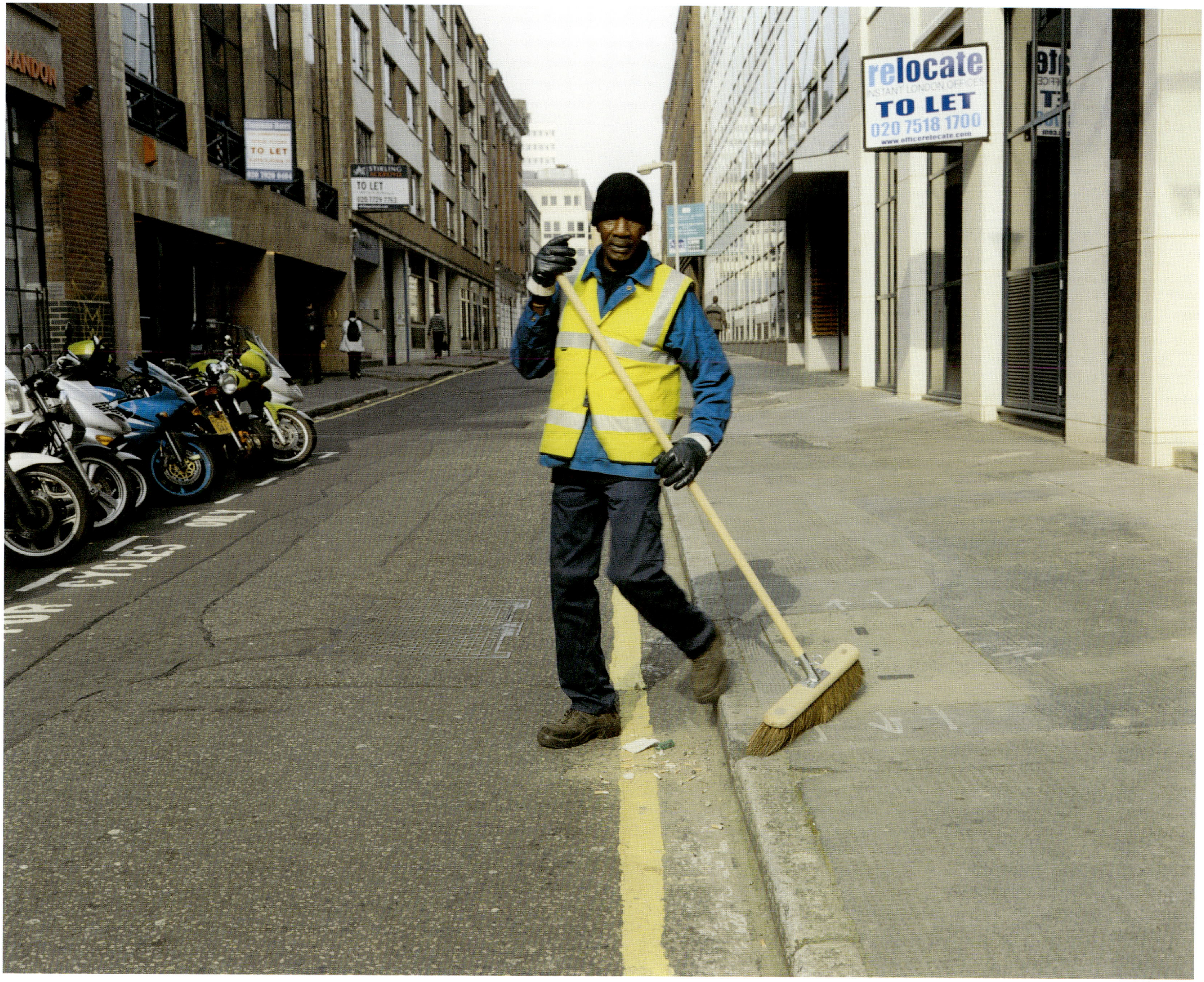
relocate
INSTANT LONDON OFFICES
TO LET
020 7518 1700
www.officerelocate.com
TO LET
TO LET
CYCLES
ONLY

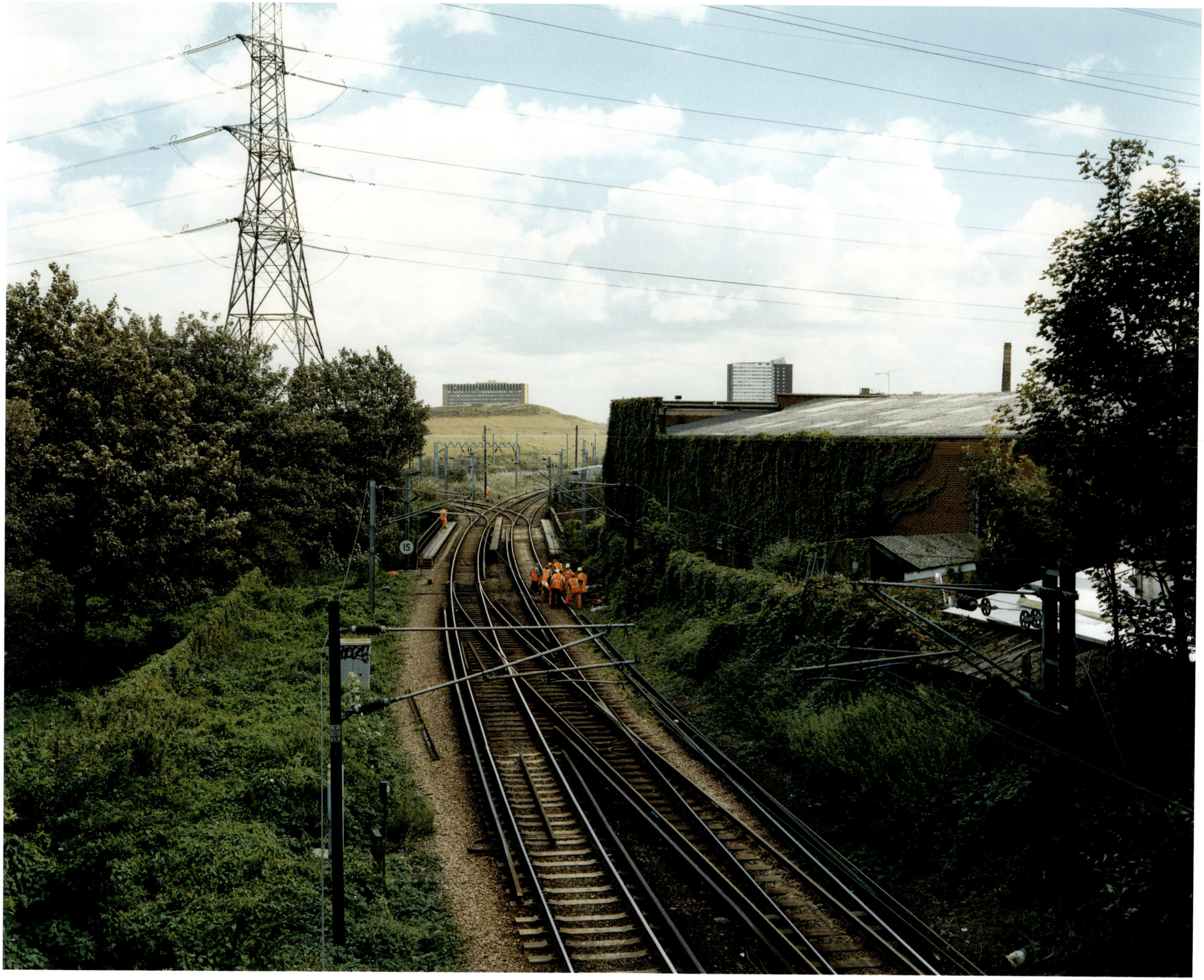

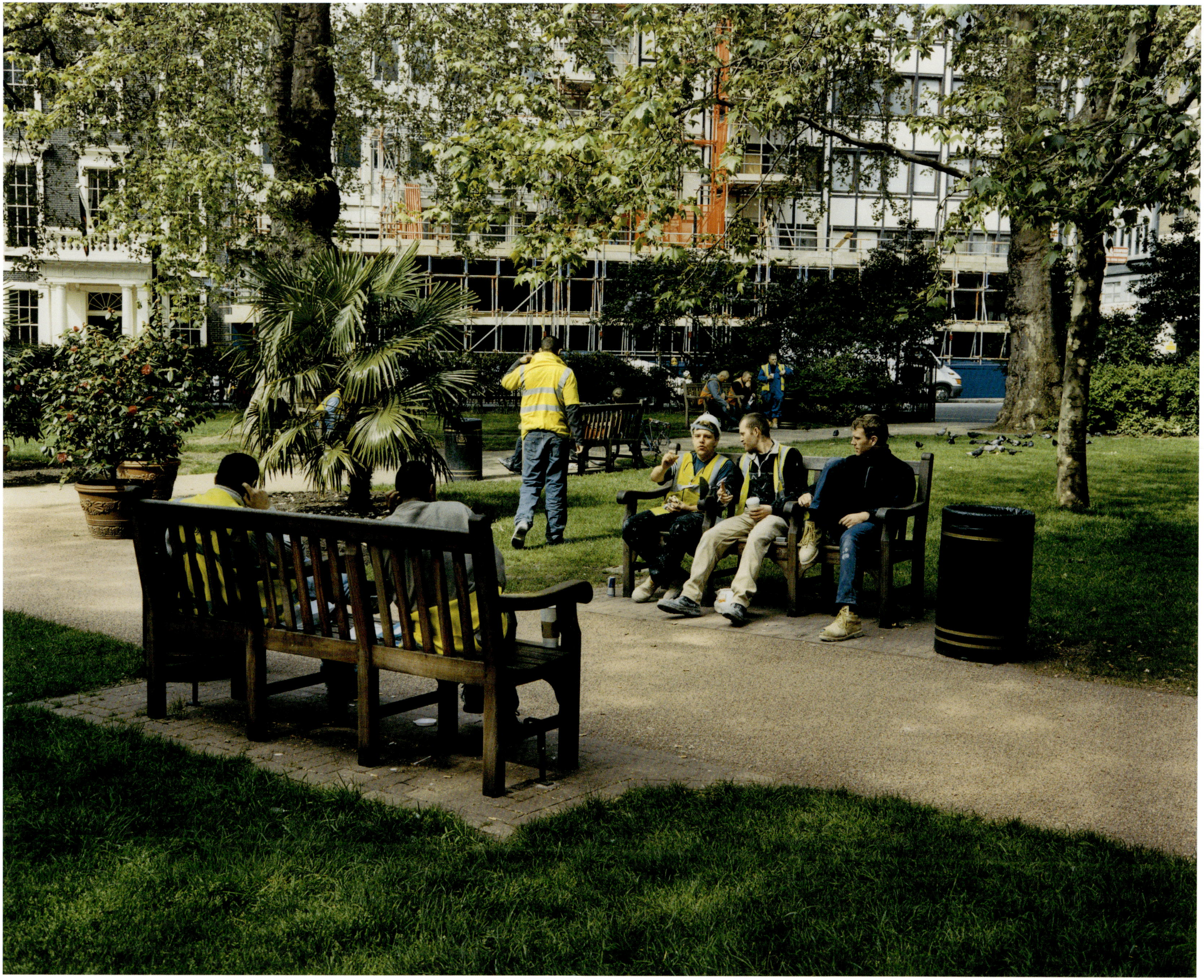

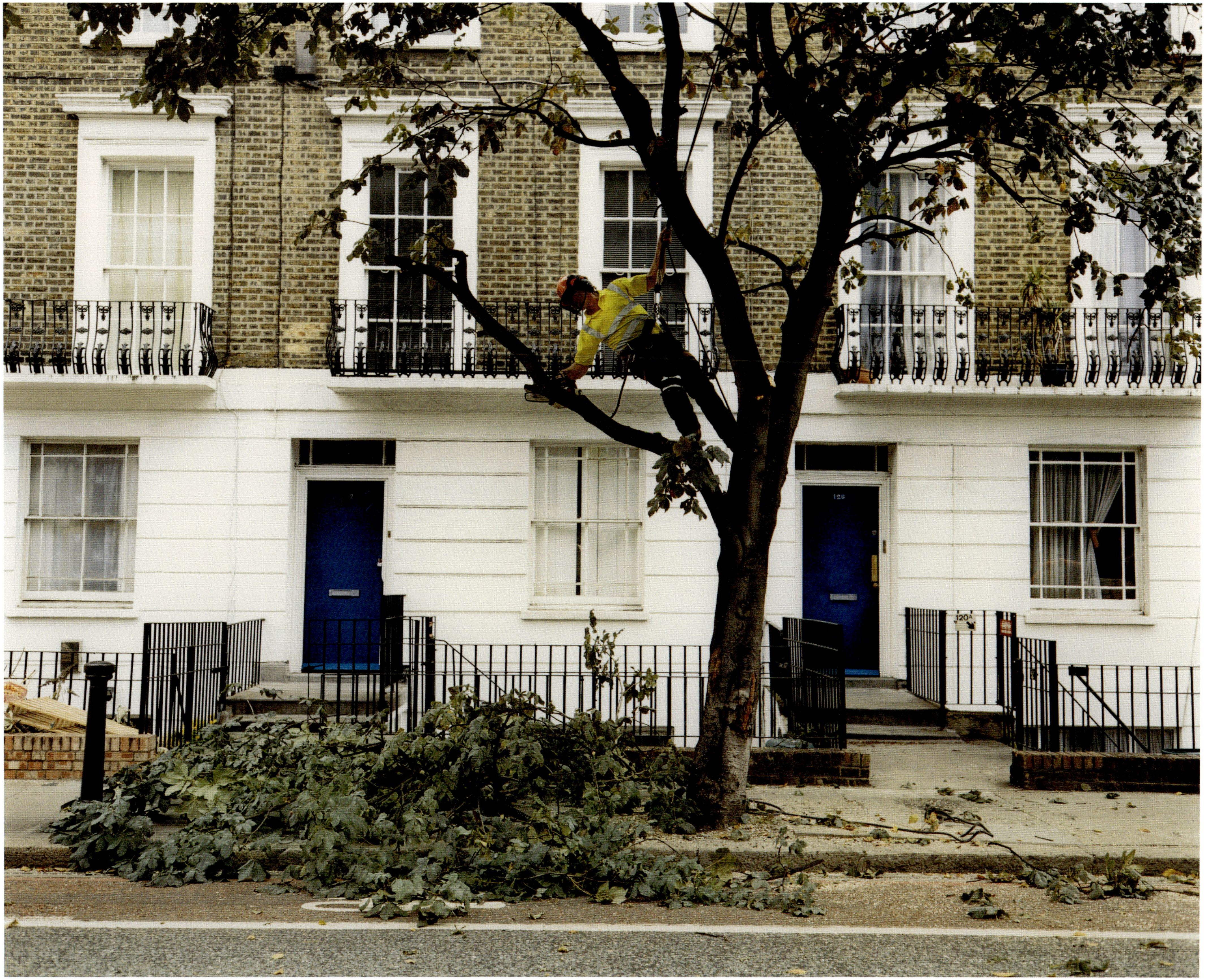

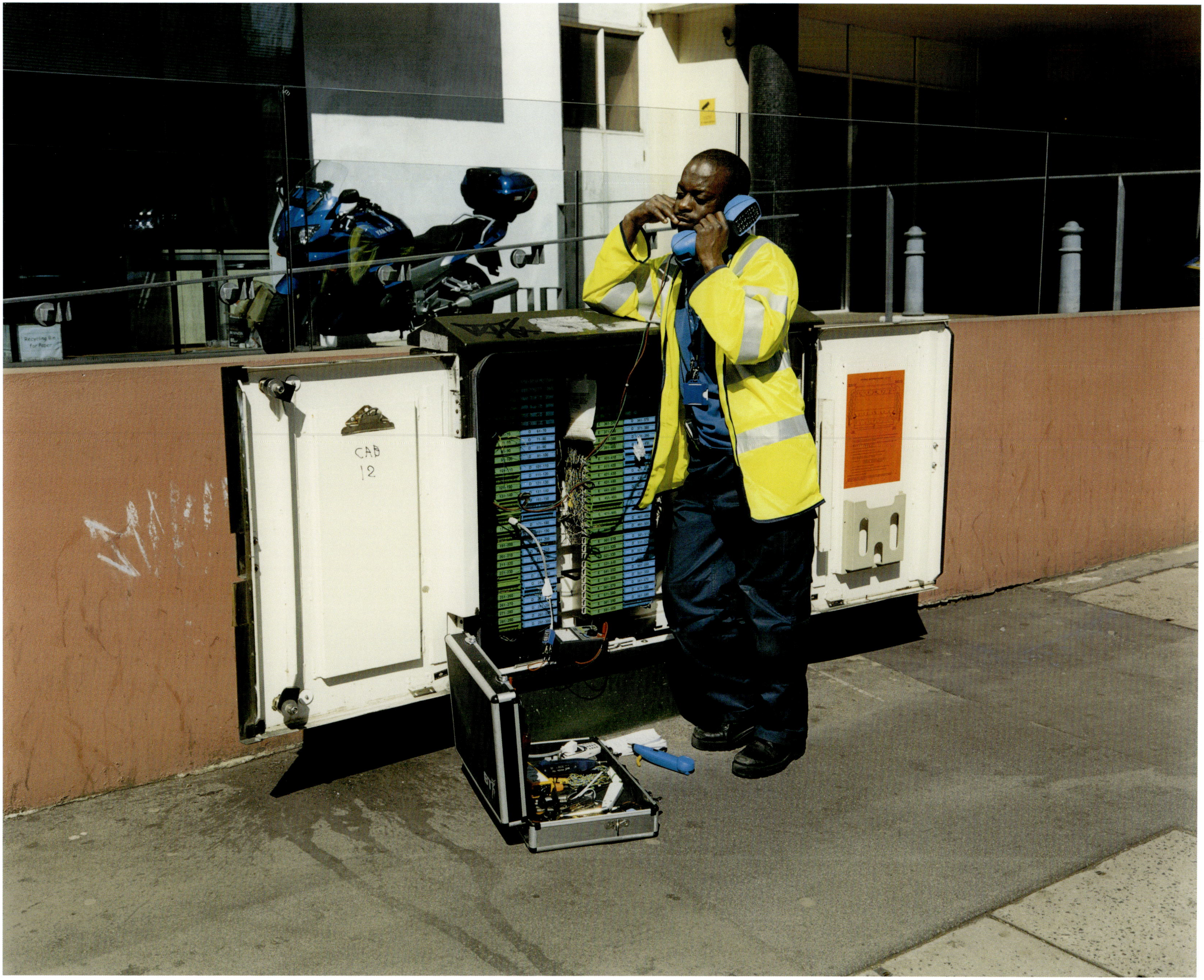
CAB
12

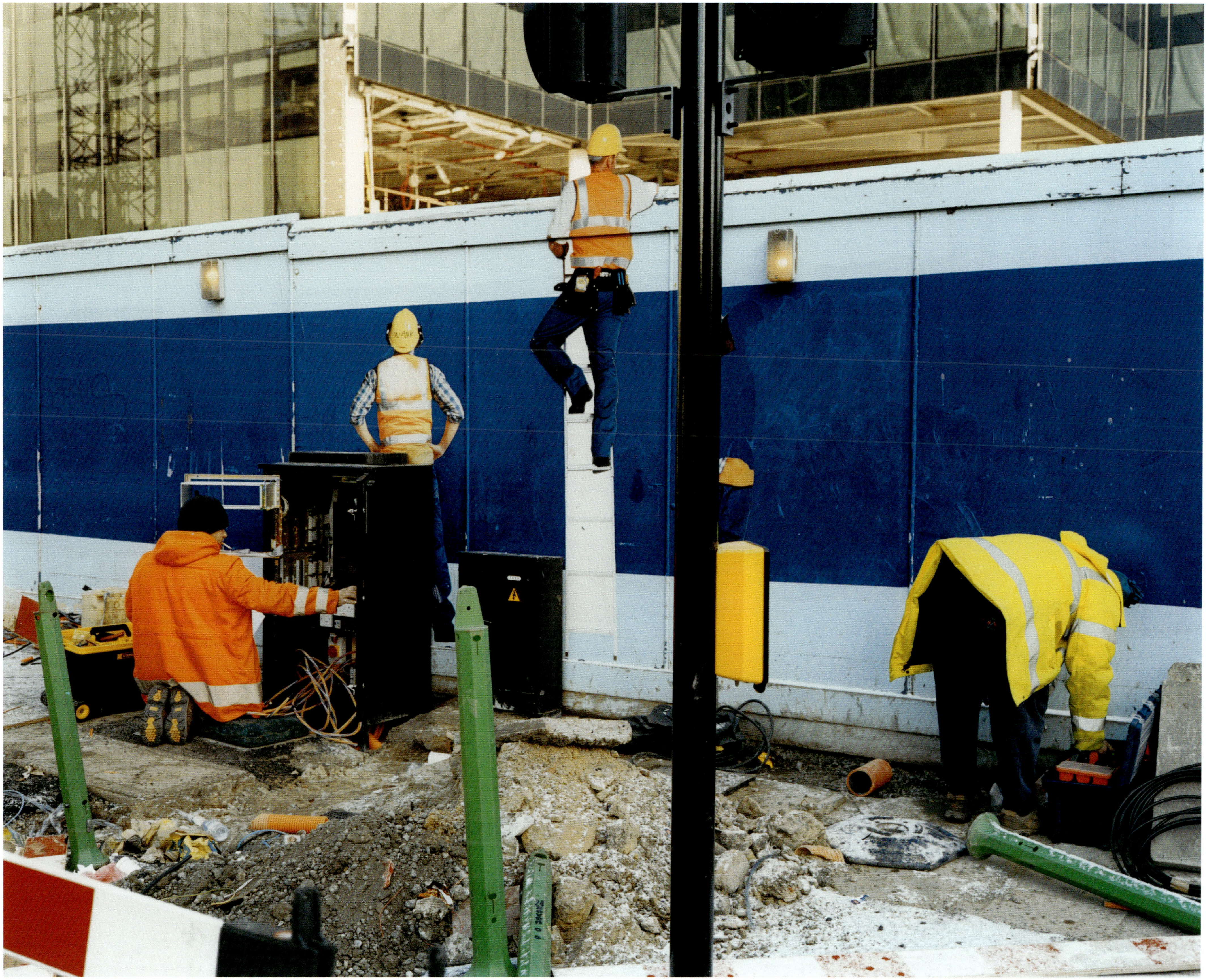

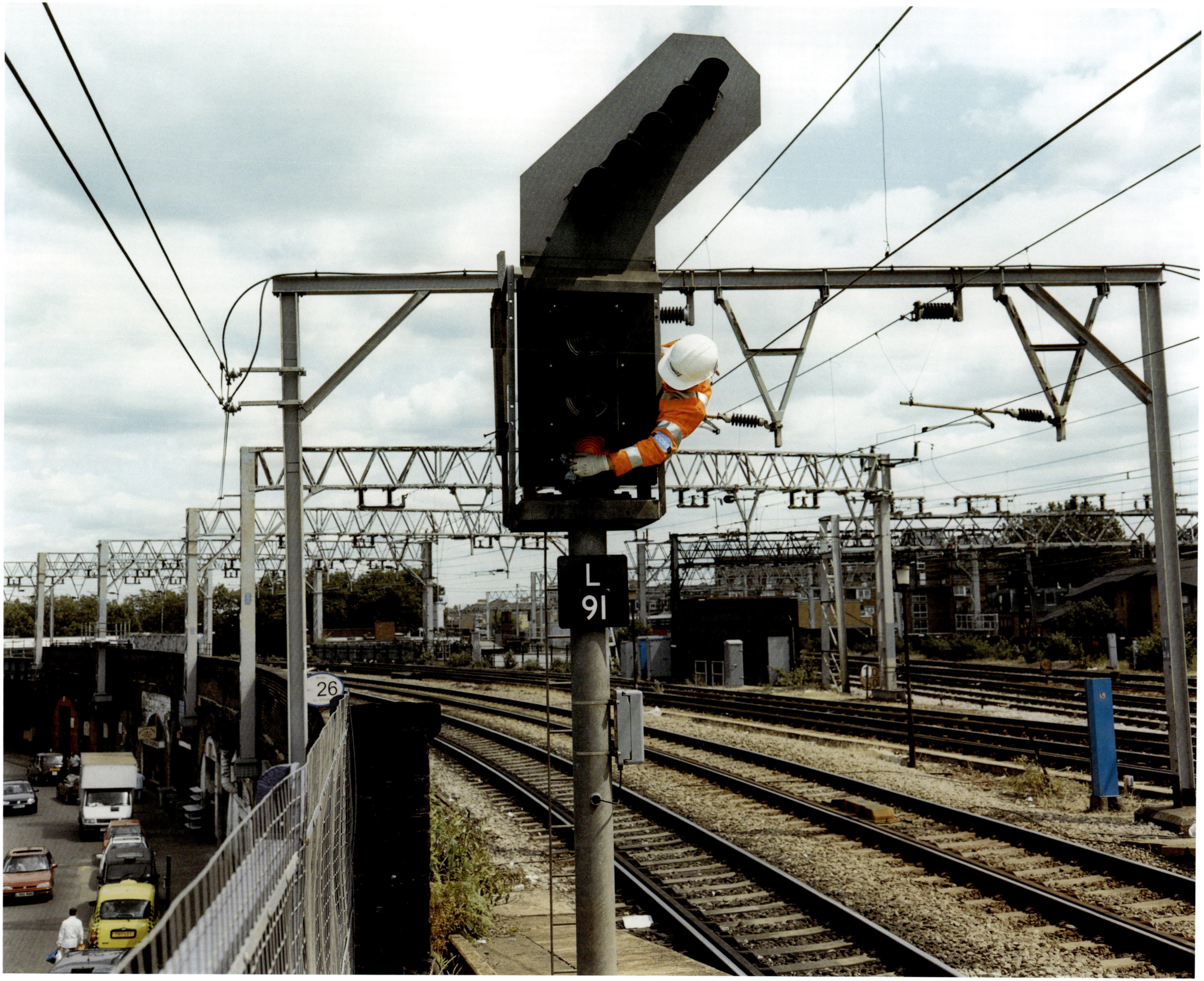
L
91
26

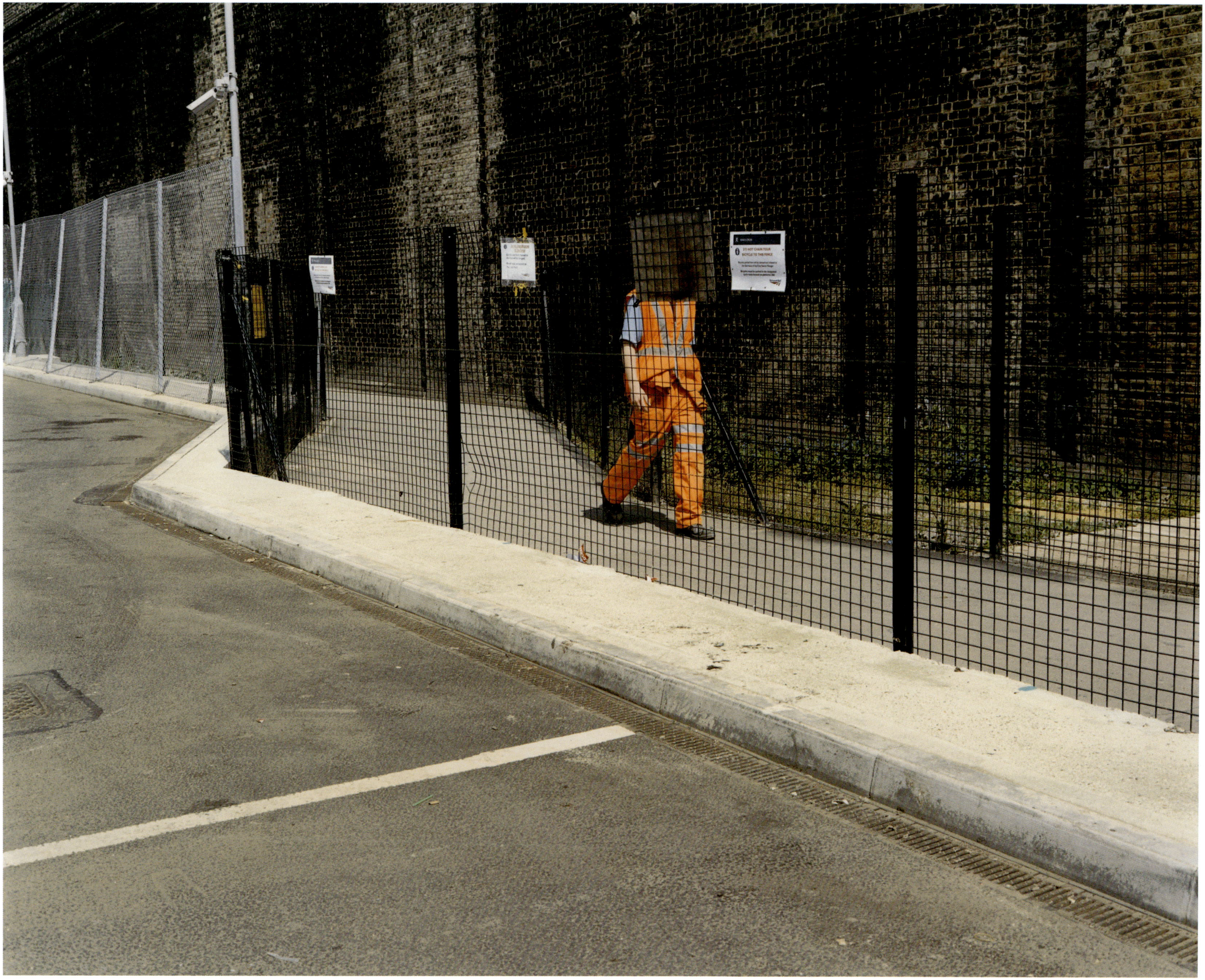
DO NOT CHAIN YOUR
BICYCLE TO THIS FENCE

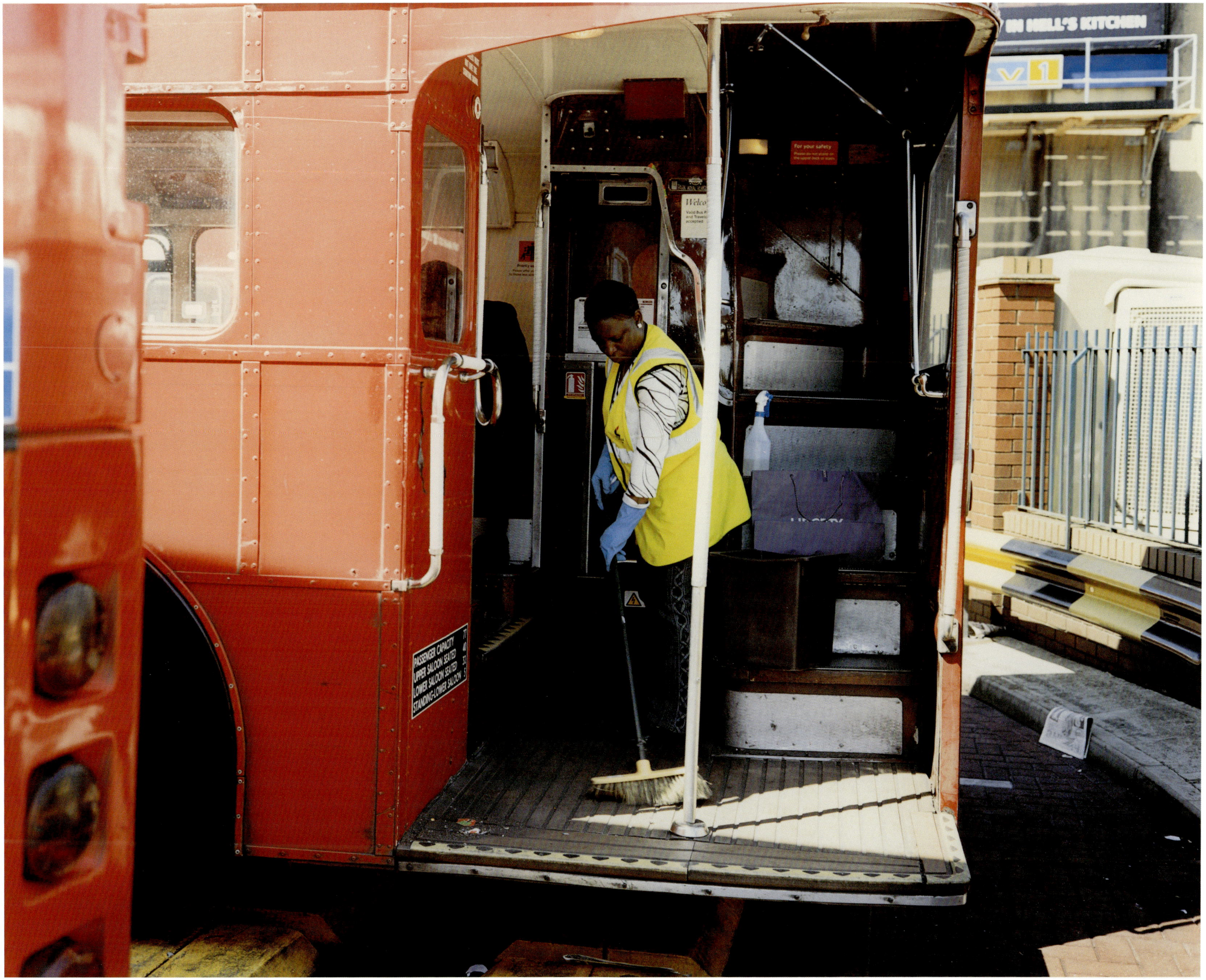
IN HELL'S KITCHEN
For your safety
PASSENGER CAPACITY 77
UPPER SALOON SEATED 40
LOWER SALOON SEATED 32
STANDING-LOWER SALOON 5

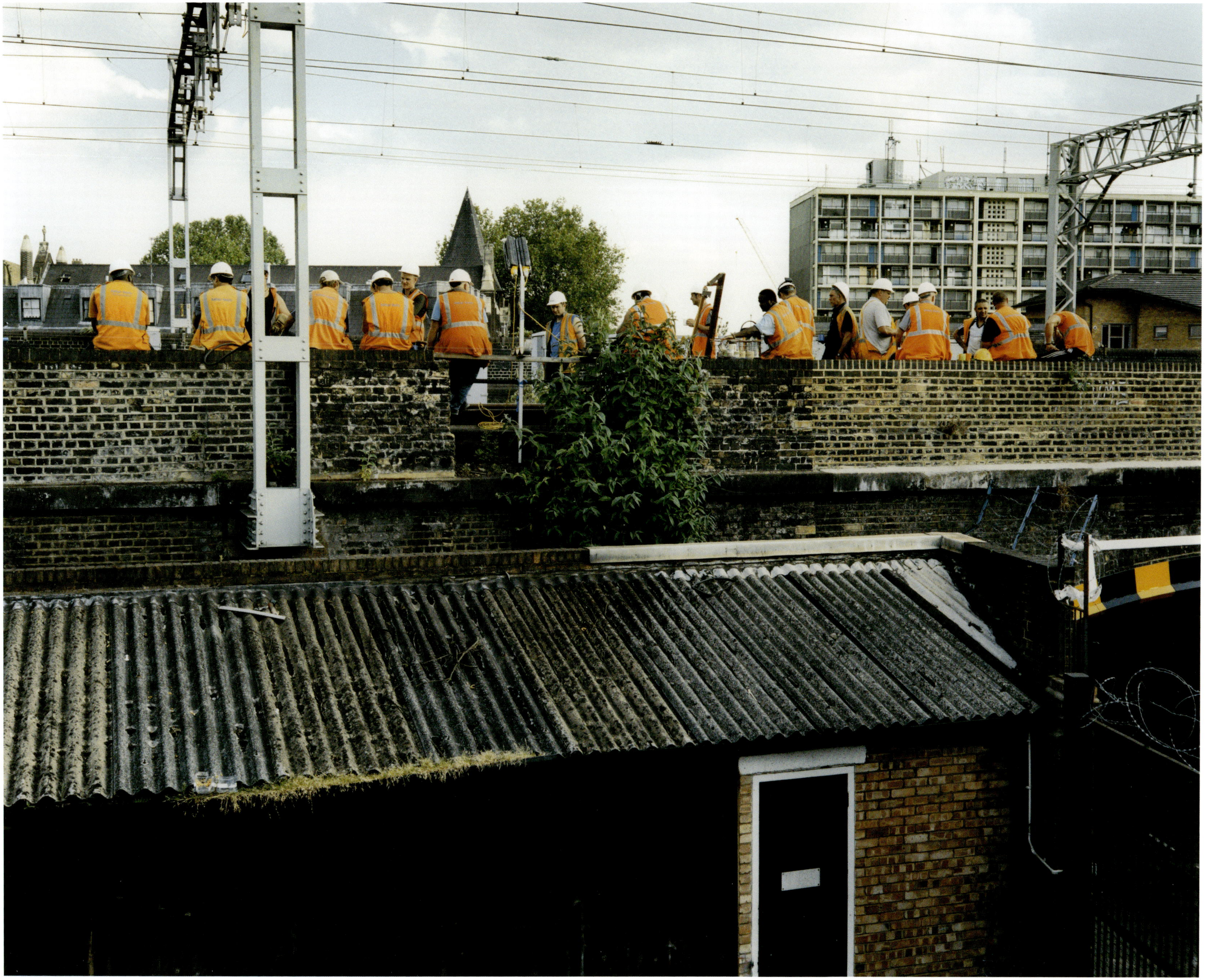

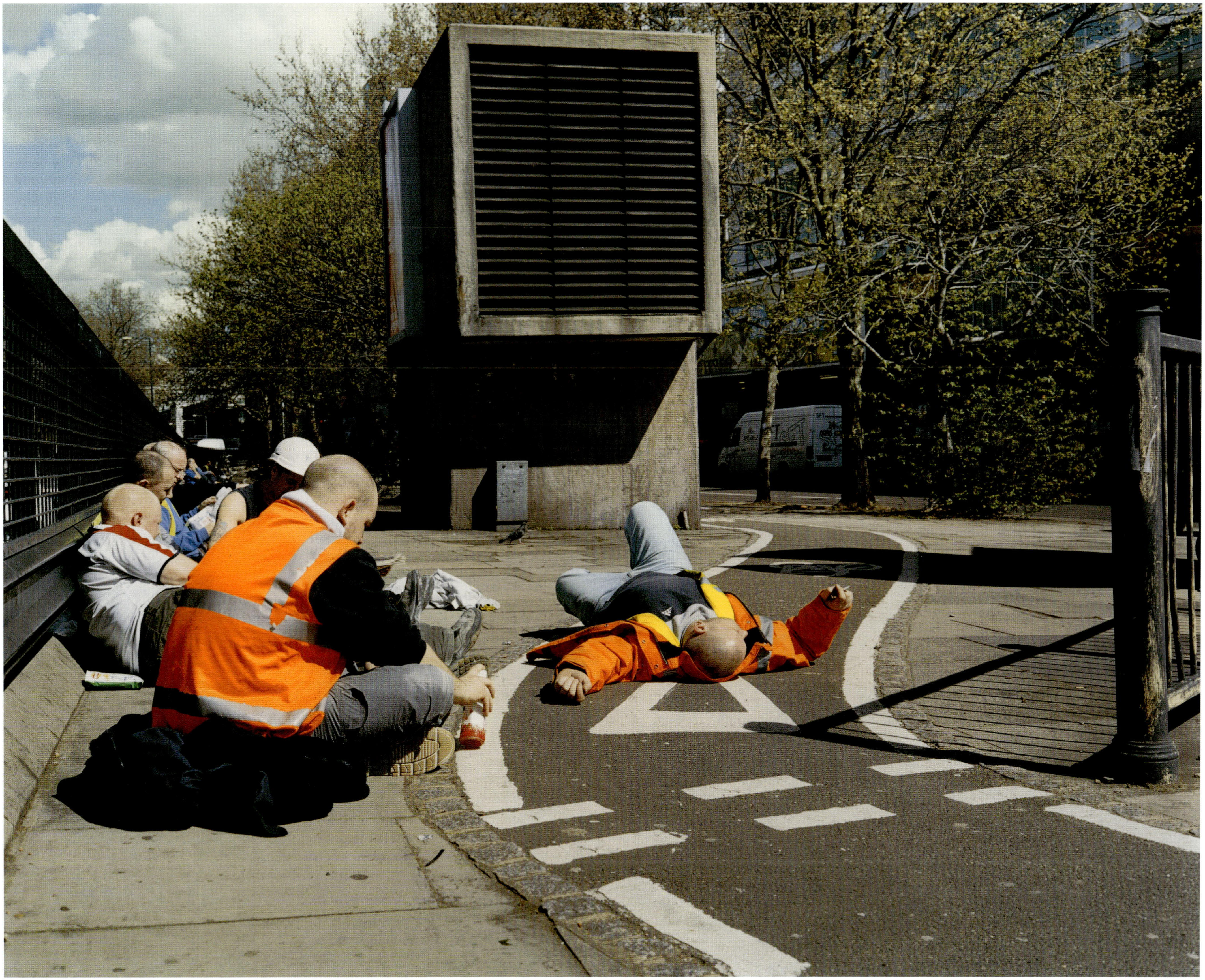

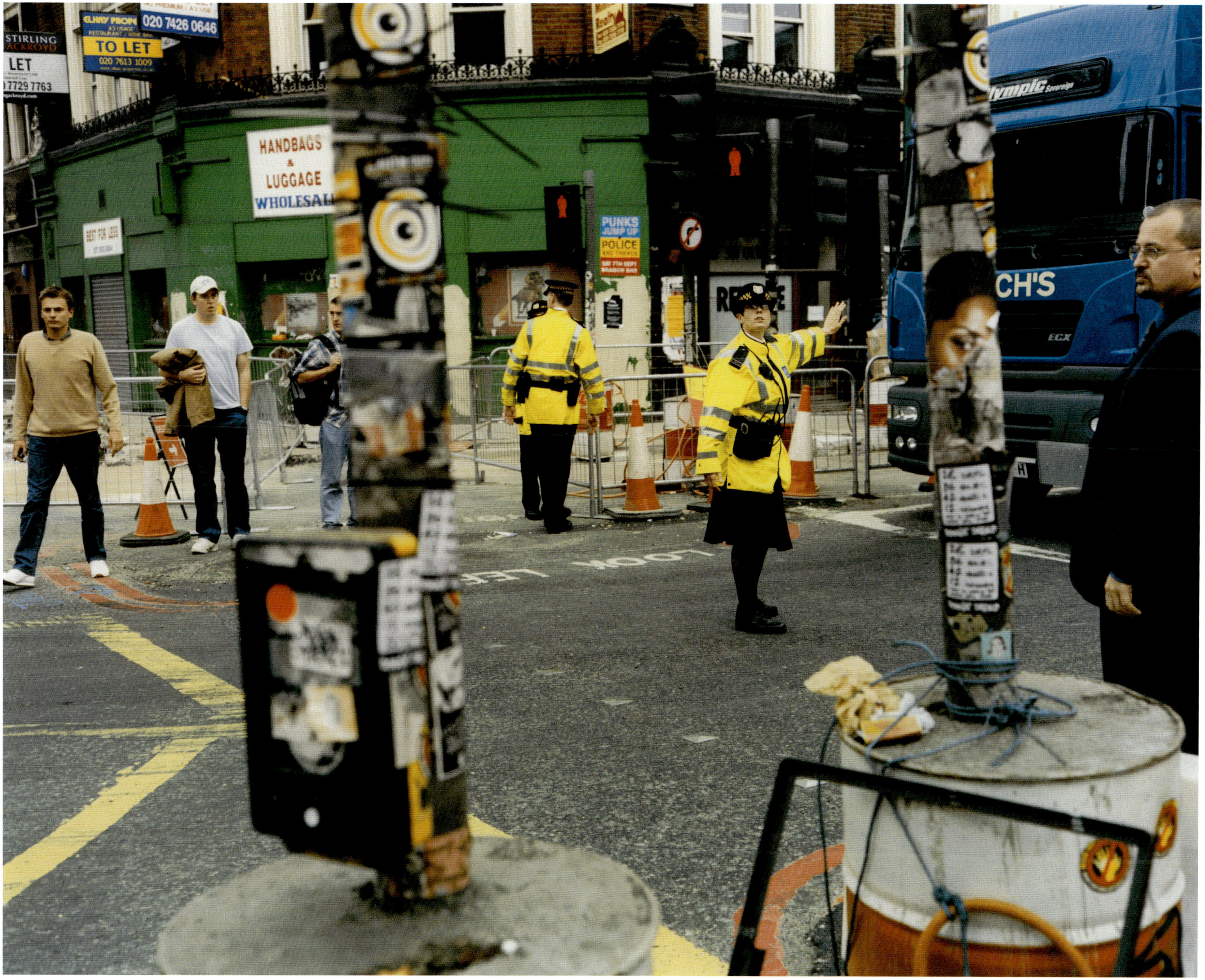

STIRLING
ACKROYD
LET
TO LET
020 7613 1009
020 7426 0646
HANDBAGS
&
LUGGAGE
WHOLESAL
BEST FOR LESS
PUNKS
JUMP UP
POLICE
CH'S
LOOK

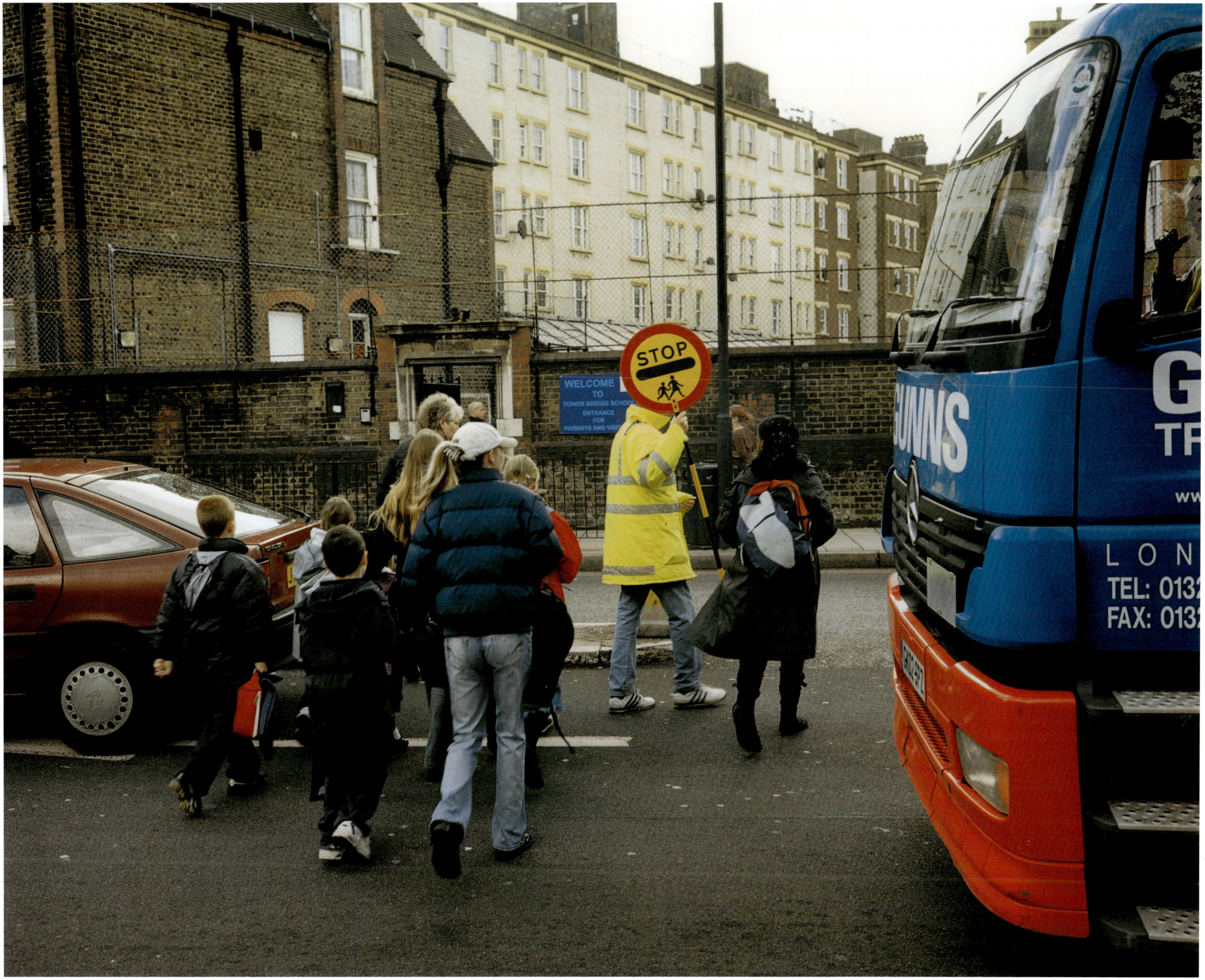
STOP
WELCOME
TO
TOWER BRIDGE SCHOOL
ENTRANCE
FOR

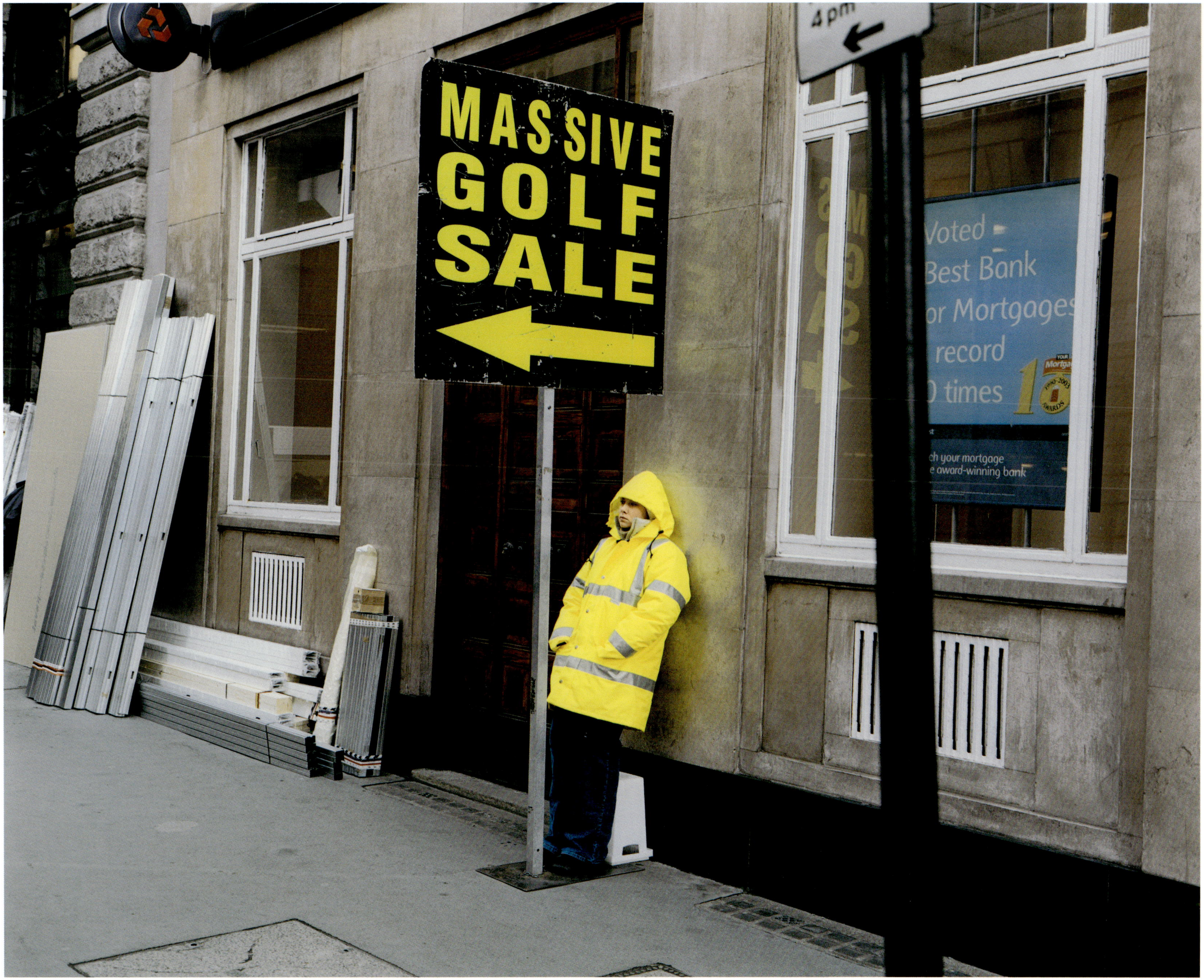
MASSIVE
GOLF
SALE
4 pm
Voted
Best Bank
record
times

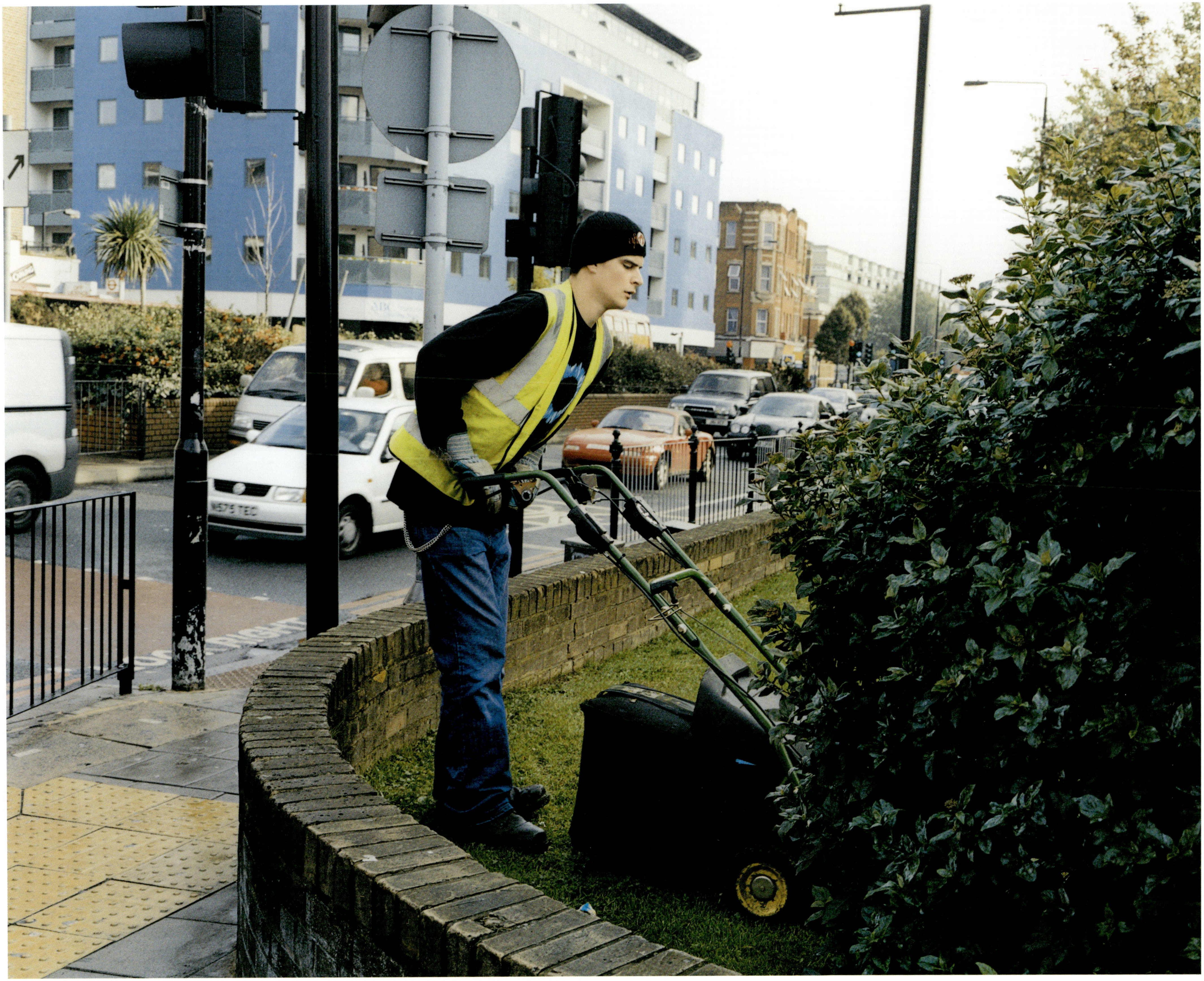

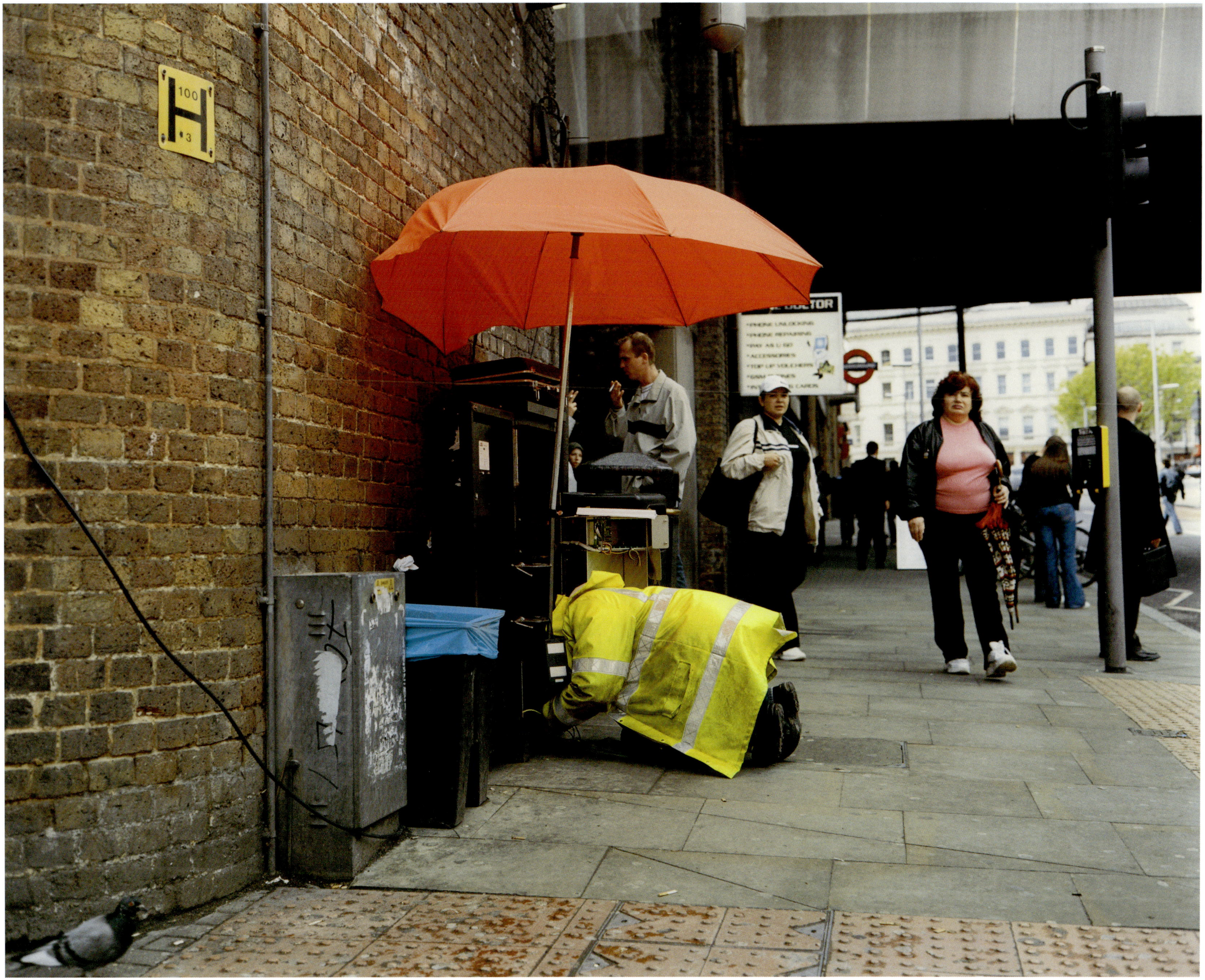
H
100
3

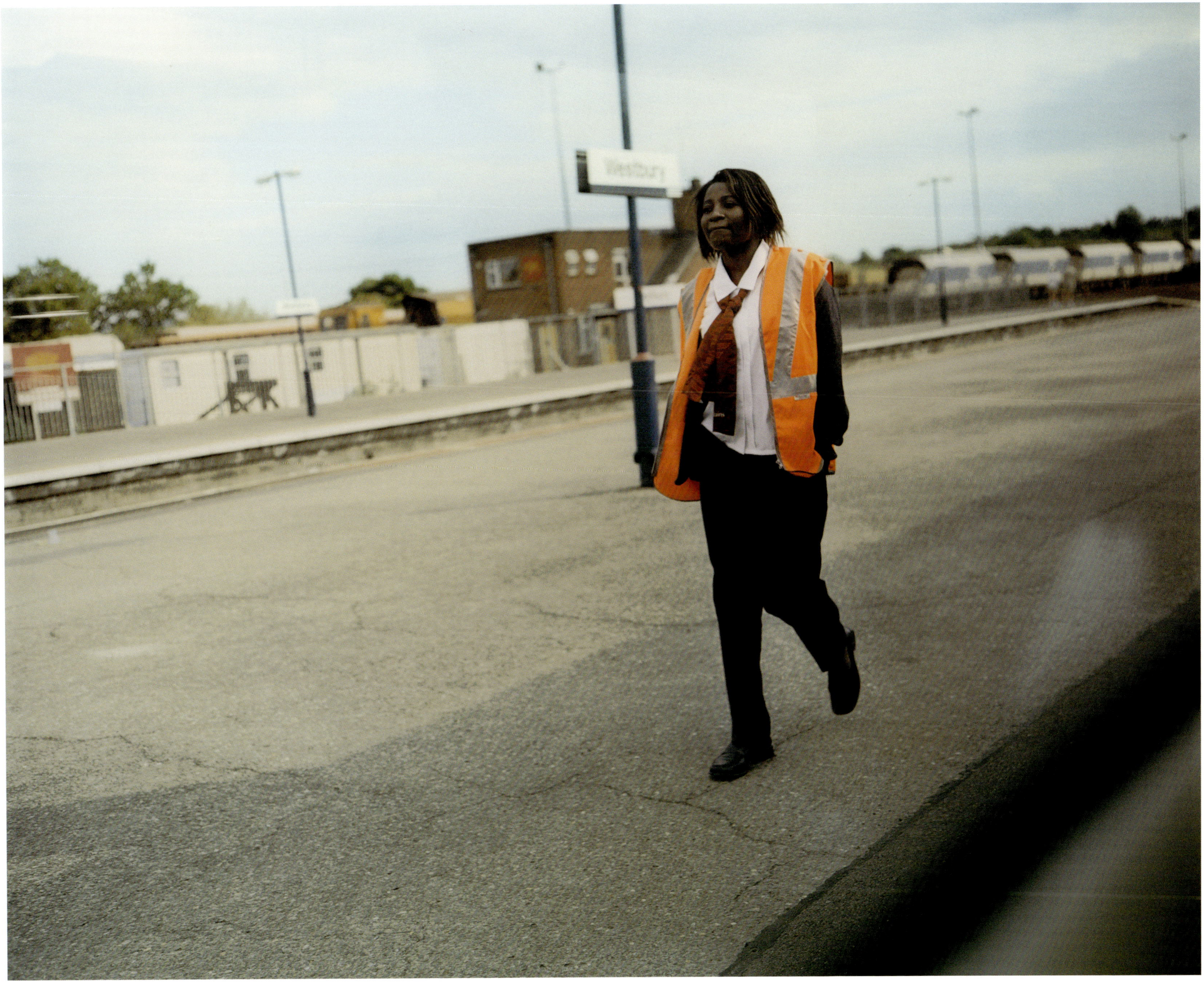

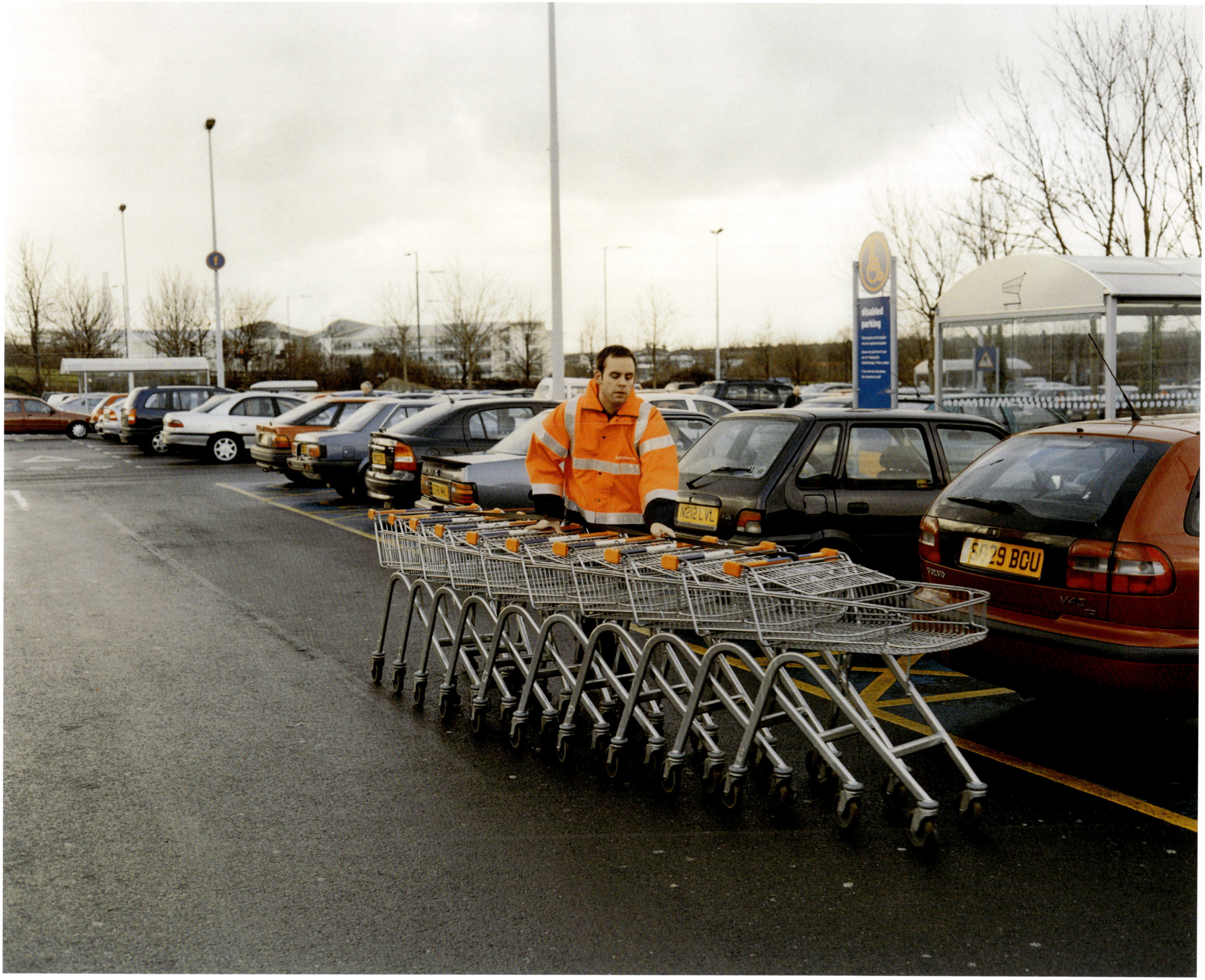
disabled parking

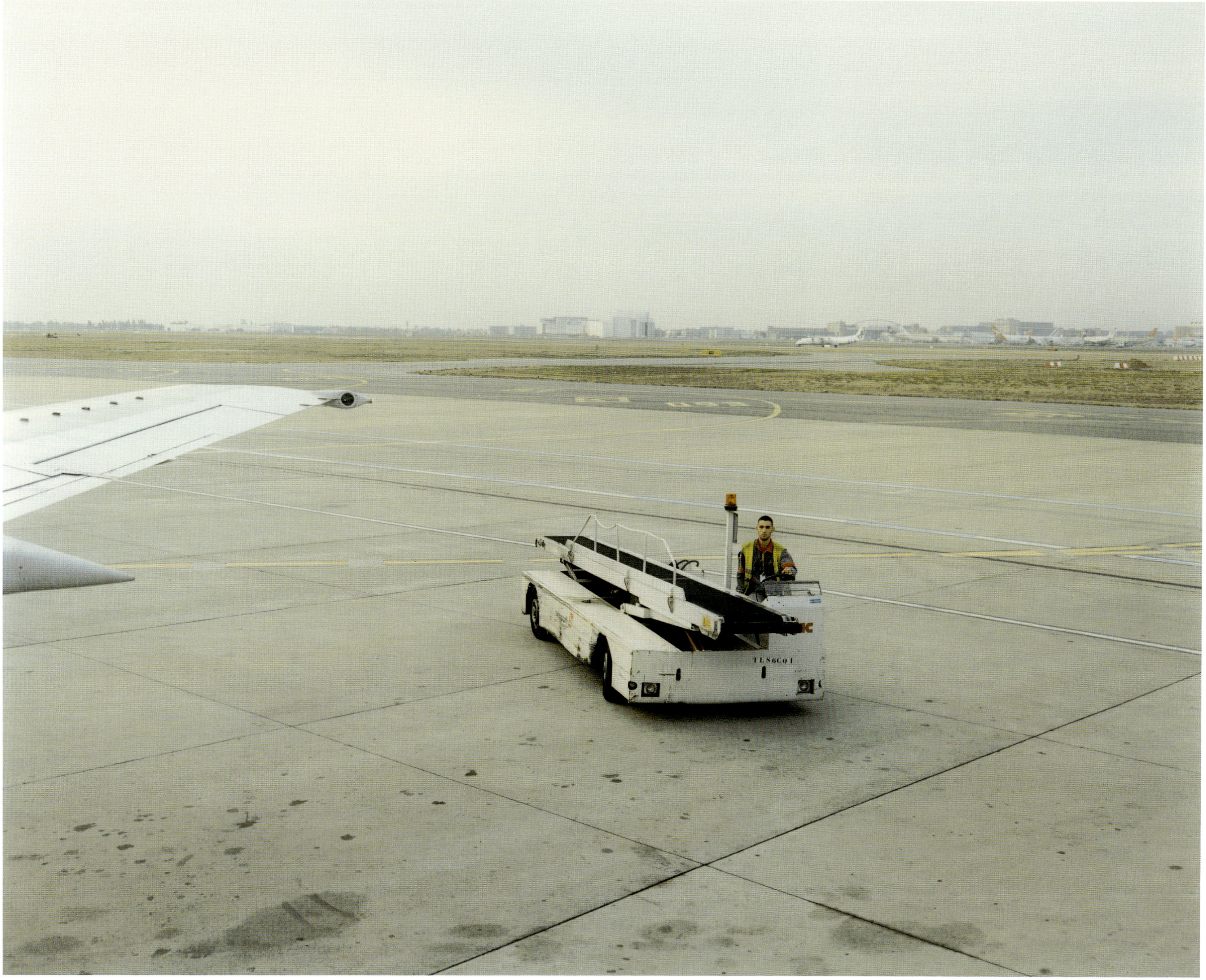
TLS6C01

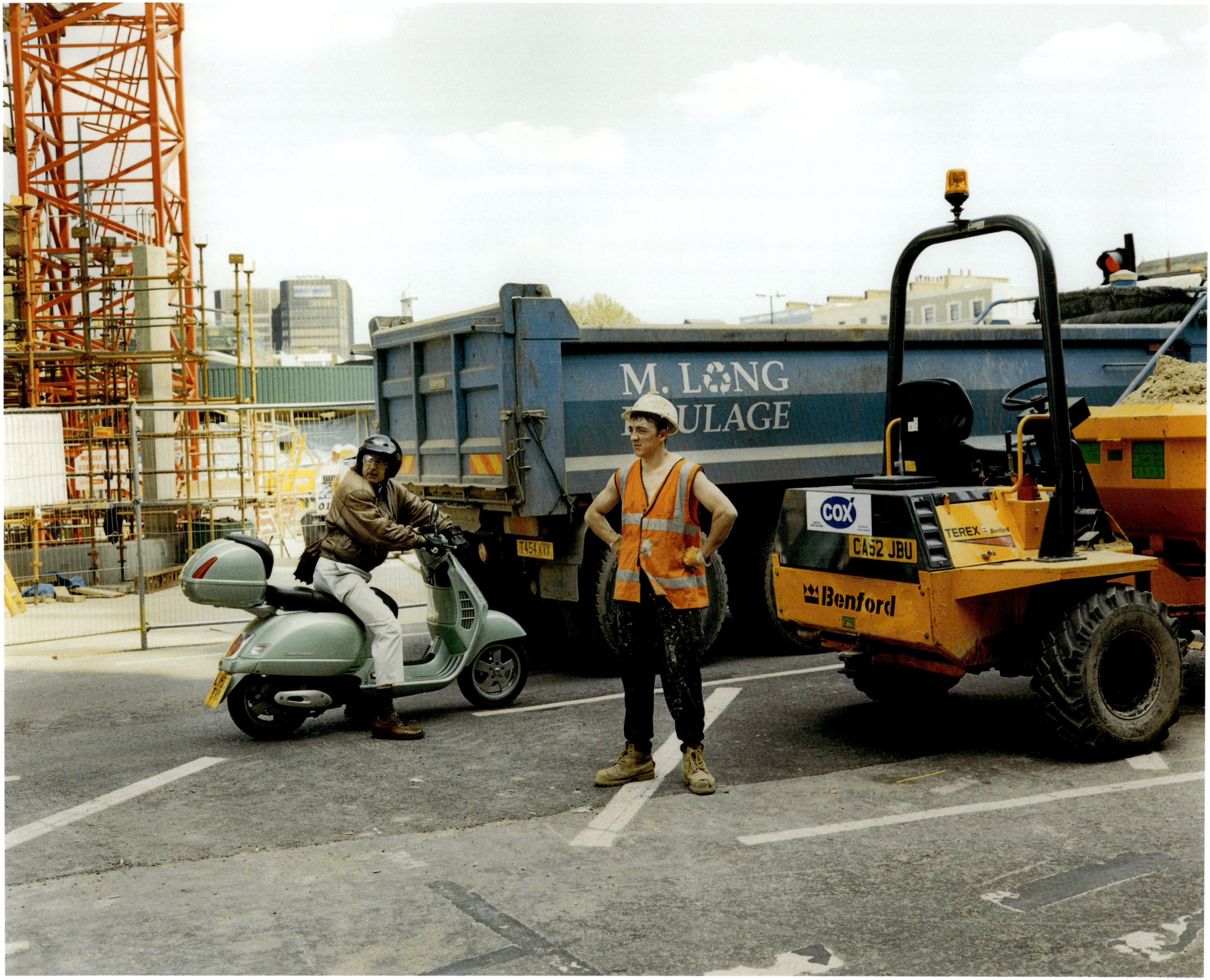
M. LONG
ULAGE
T454 KVX
COX
CA52 JBU
TEREX
Benford

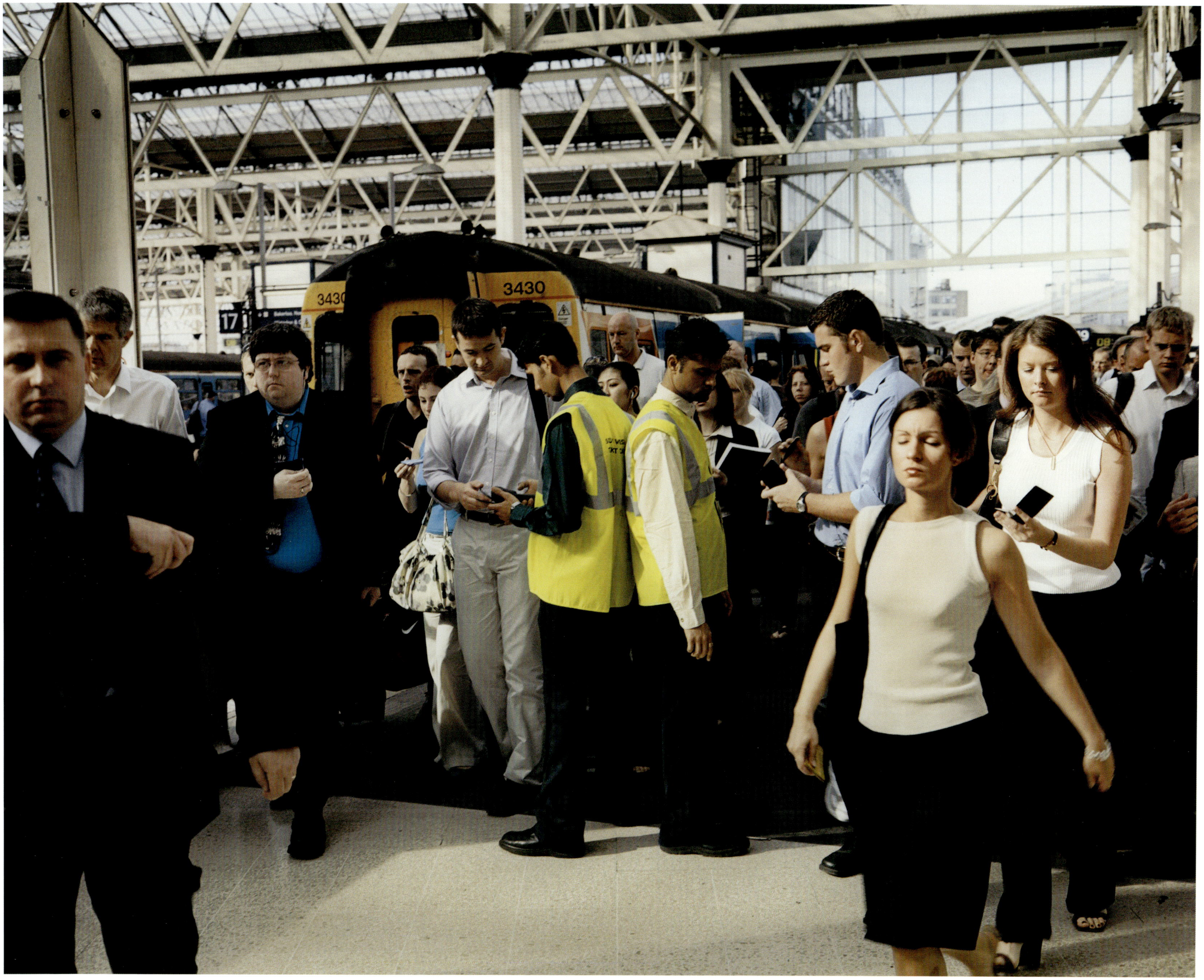
3430
3430
17

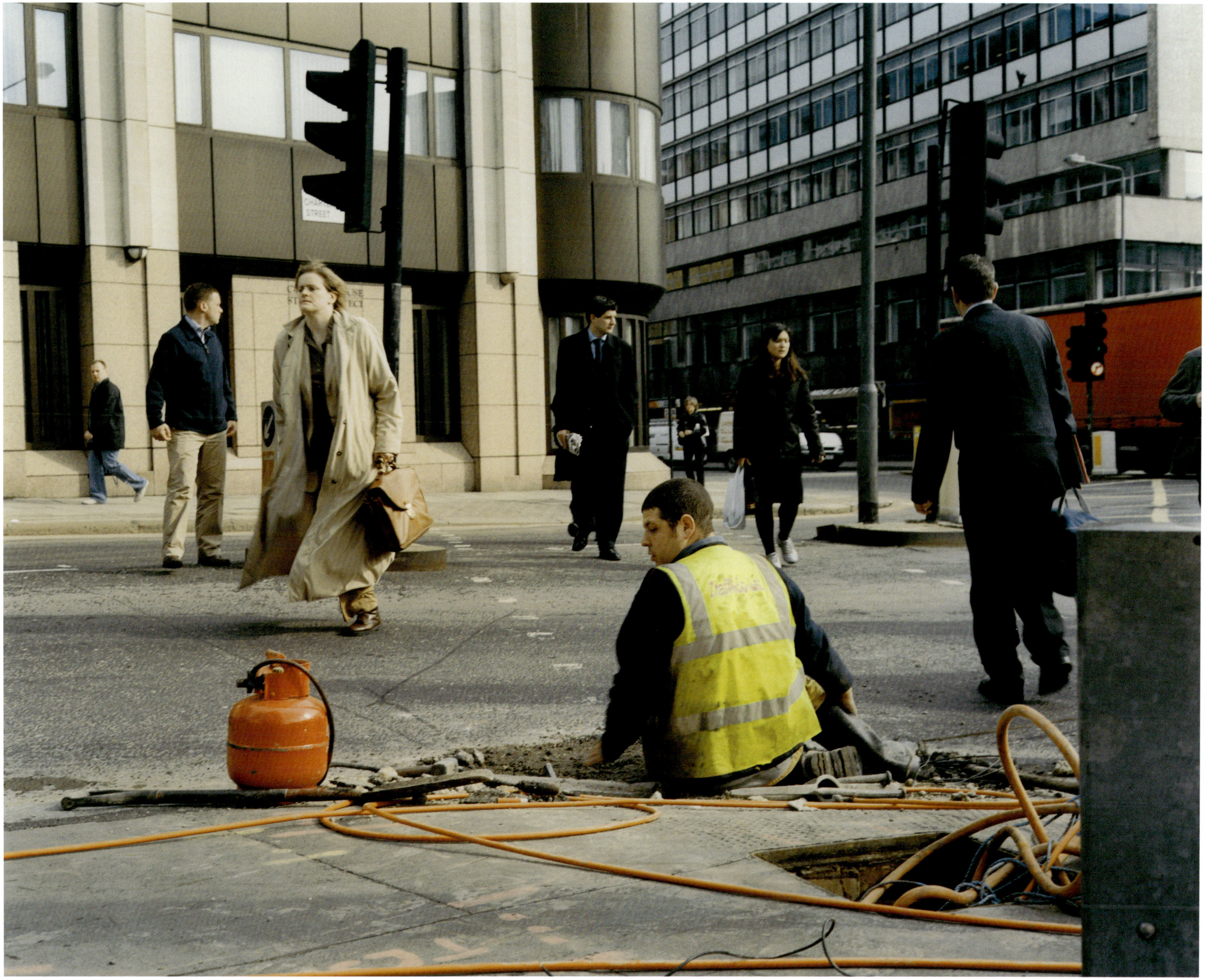
STREET

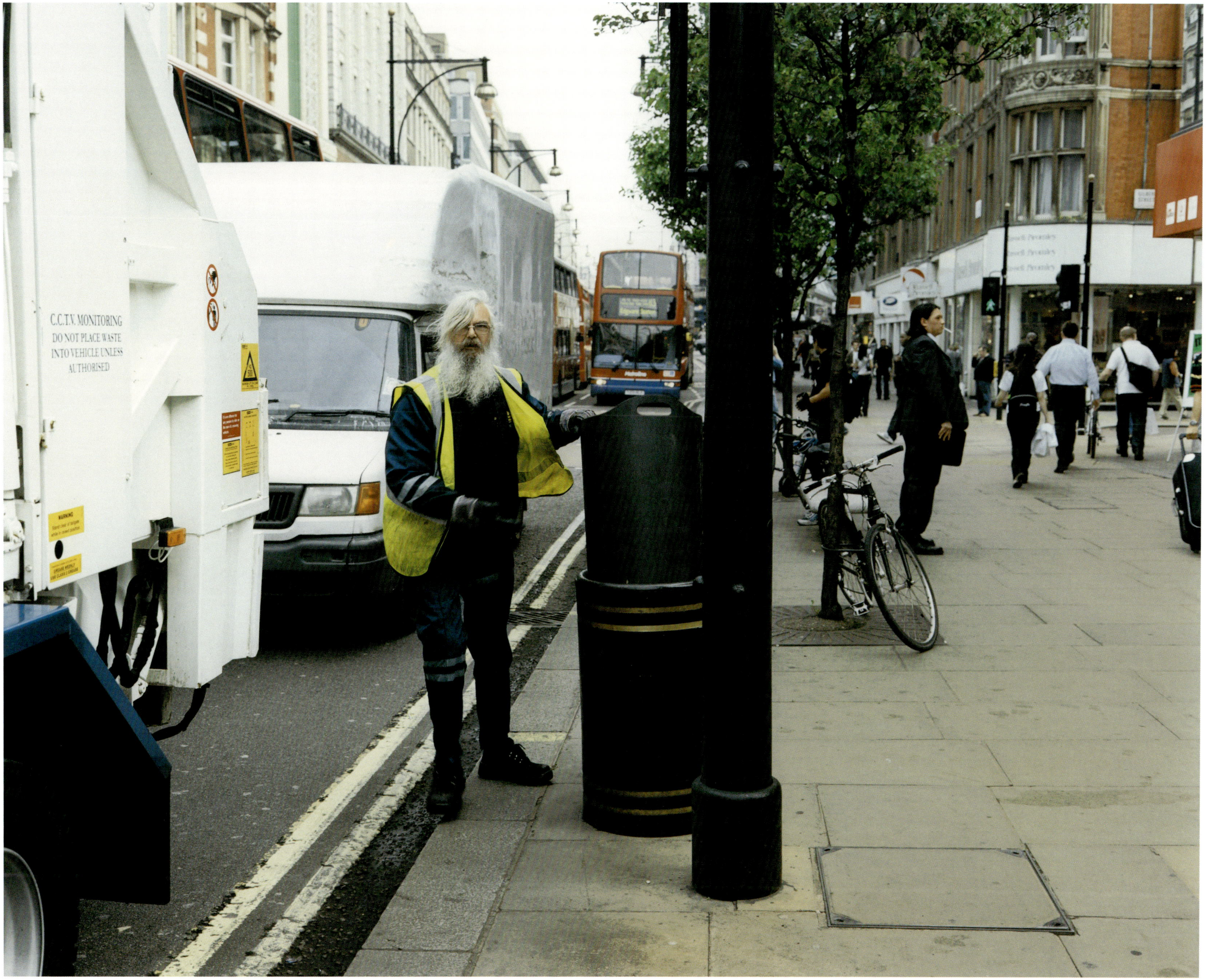
C.C.T.V. MONITORING
DO NOT PLACE WASTE
INTO VEHICLE UNLESS
AUTHORISED

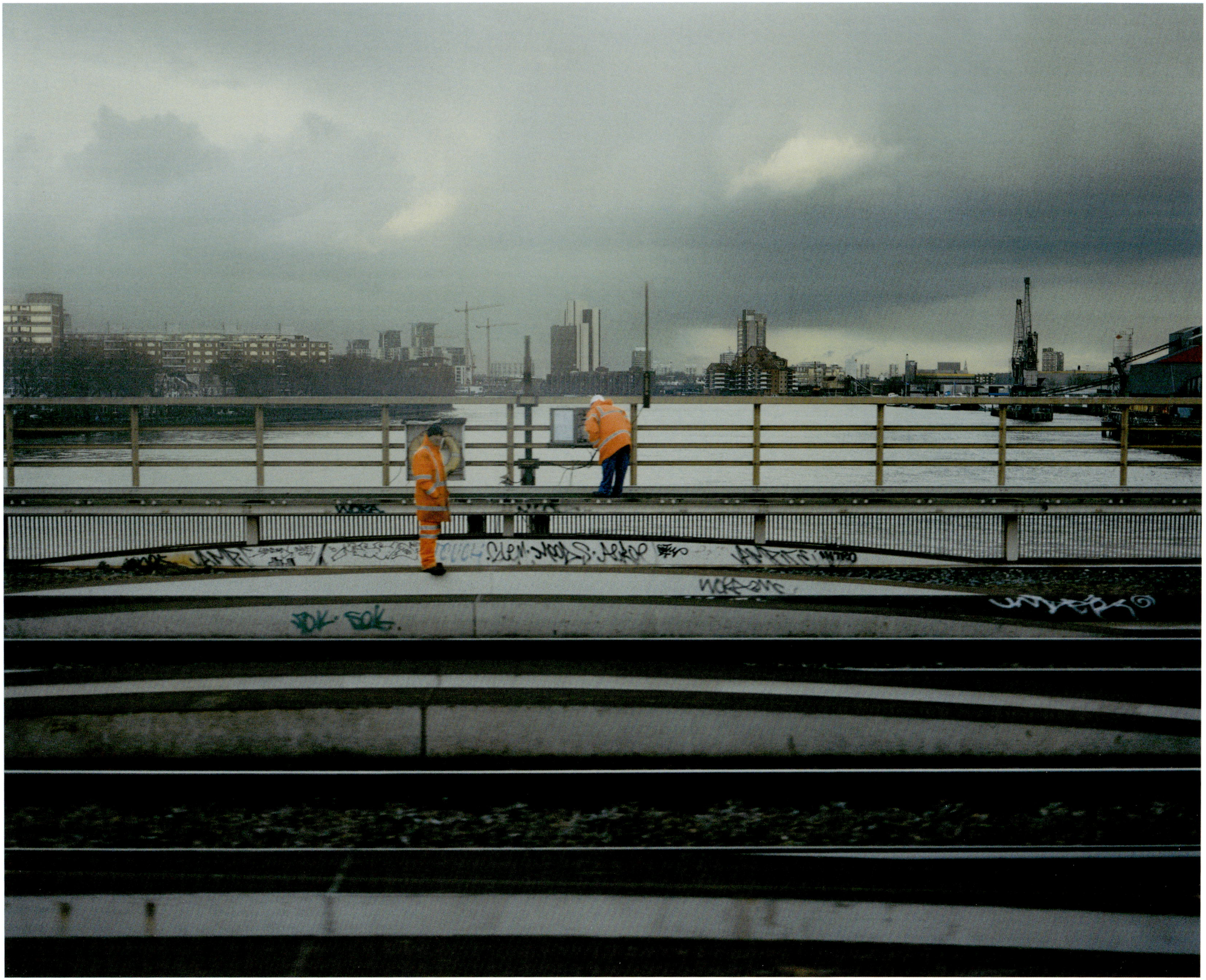

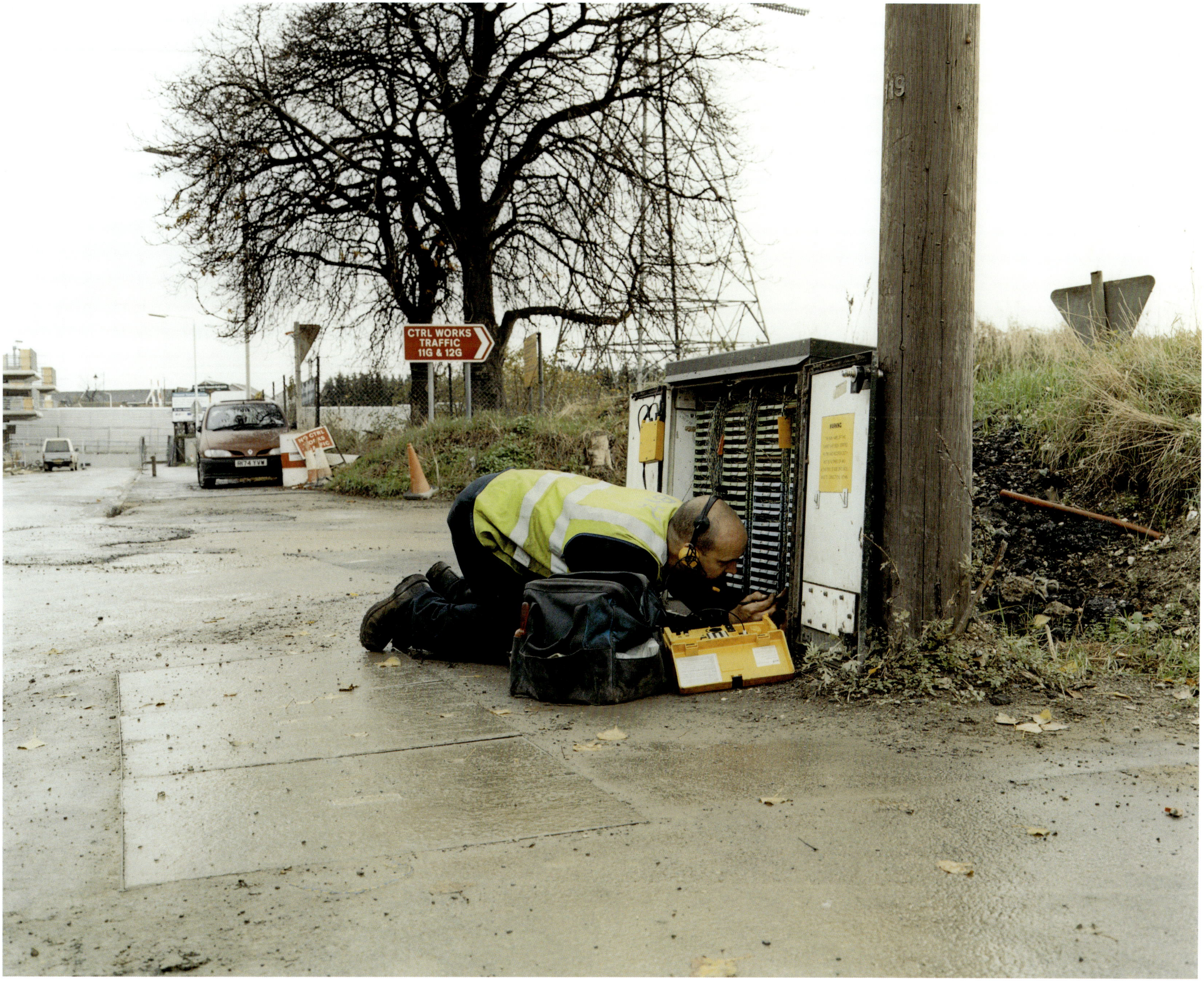
CTRL WORKS
TRAFFIC
11G & 12G

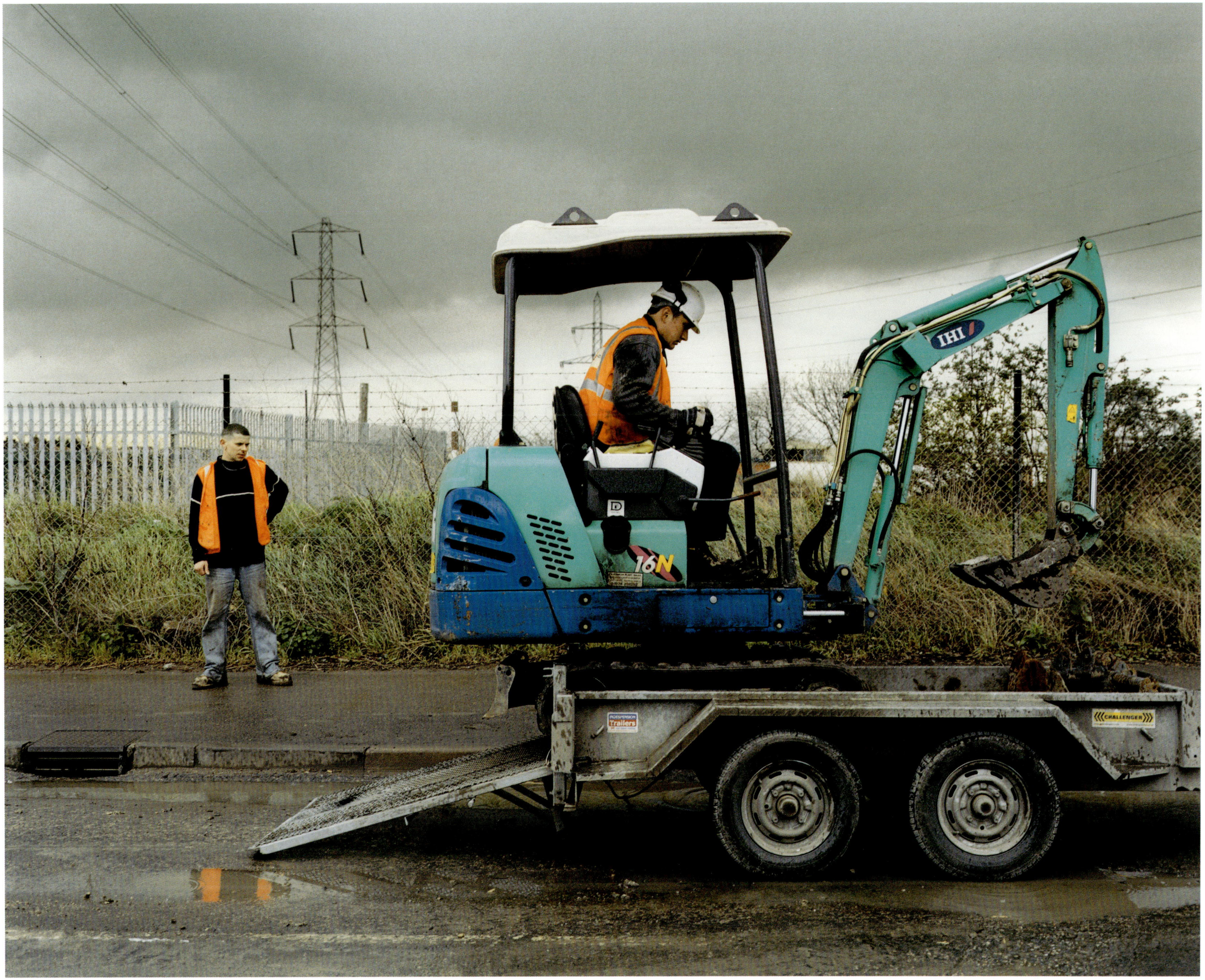
IHI
16N
Trailers
CHALLENGER

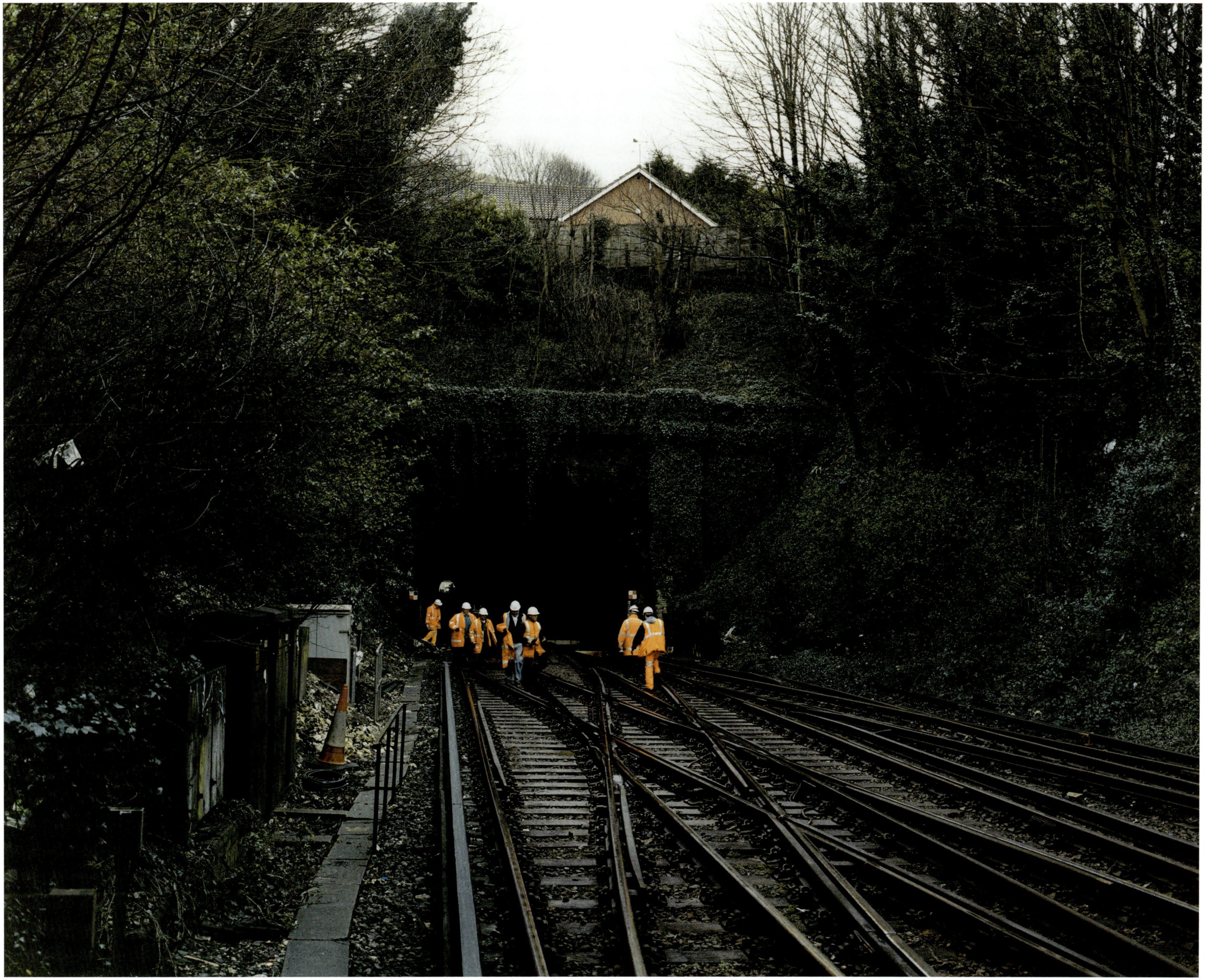

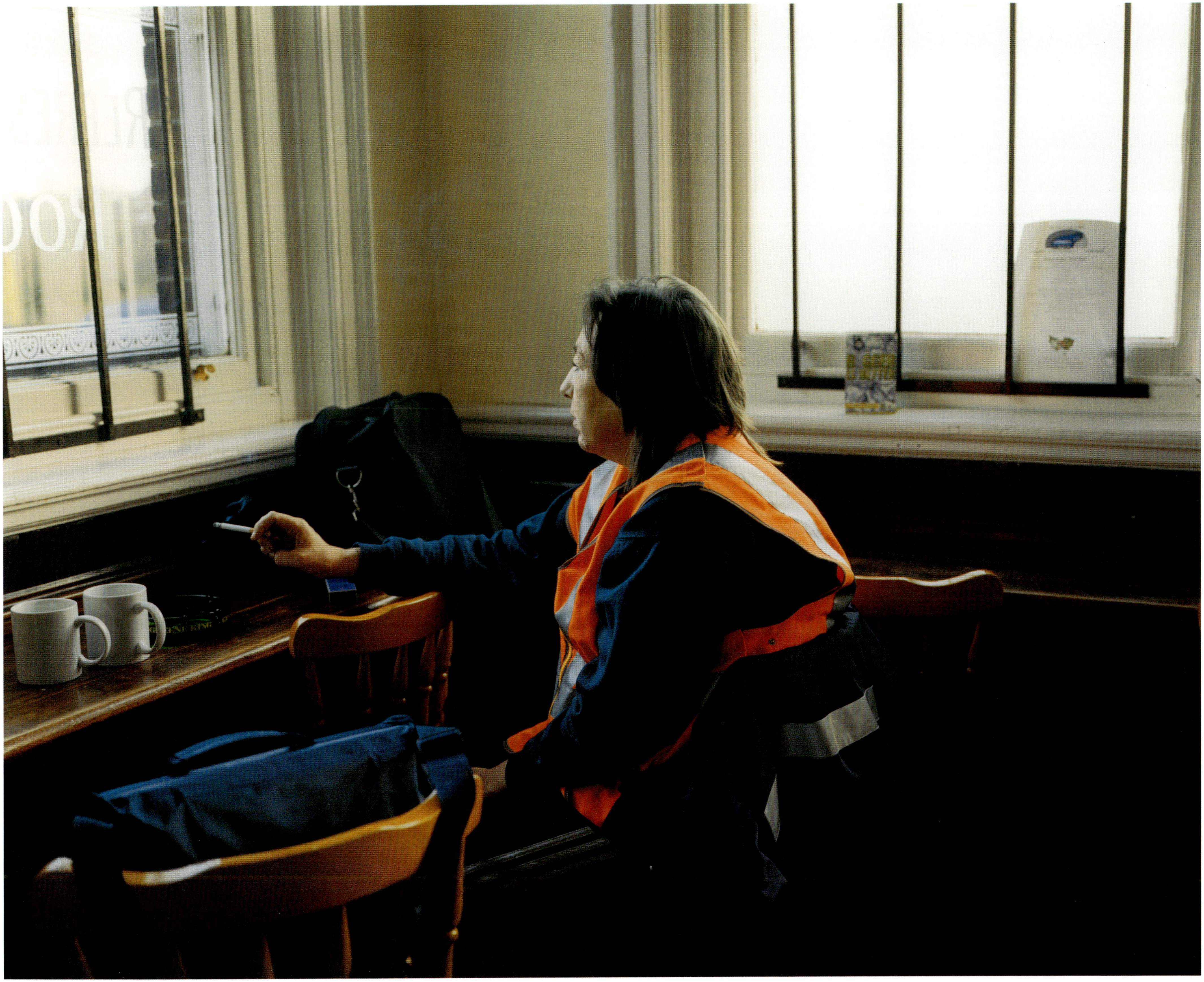

Invisible by Stephen Gill

First edition, 1000 copies
Published 2005 by Nobody


www.stephengill.co.uk

This book was published on the
occasion of an exhibition at the
Victoria Miro Gallery, London.

Book design by Melanie Mues
Preface by Mark Haworth-Booth
C type prints by Rob Sara
Edited by Stephen Gill
Printed in Italy by EBS



Mark Haworth-Booth is Honorary
Research Fellow, Victoria and Albert
Museum and Visiting Professor
of Photography, University of the
Arts London.

ISBN 0-9549405-0-4